Essentials of Plant Pathology

Essentials of Plant Pathology

Obasai Nwosu

2026

Agri Horti Press

New Delhi-110 084

ISBN 9789383285839 (HB)

Published By

Agri Horti Press

New Delhi

E-mail: agrihortipress@gmail.com

Digitally Printed at : Replika Press Pvt. Ltd.

Preface

Plant pathology is warmly connected to the health of plant and the invention of crop yields. It has special significance in agricultural countries like India which are fast developing and where much still needs to be done in the renewal of organize measures against plant diseases. Rough guesses have suggested that in India, on an average, there is a loss of about 10 percent in crop yield due to plant diseases. This loss would surely amount to hundreds of crores of rupees annually, taking into account the output of all the refined crops. Hence, a on plant pathology is always welcome.

The book has numerous significant features, one of which is the "all-in-one" approach which will be useful to students. To the best of my knowledge there is no book with both generalized and specialized topics discussed in parallel which caters to the needs of undergraduate, postgraduate and research students at the same time. Most of the important aspects of plant pathology, such as physiology of diseased plants, root diseases, send-born diseases, post harvest diseases of fruits and vegetables, diseases caused by fungi, bacteria, viruses, mycoplasmas, nematodes and non-parasitic diseases have all been dealt with suitable detail for the benefit of students. Besides, the book should also serve as a good reference of research workers.

It is hoped that this work will enrich the background of those who wish to study plant pathology intensively. Suggestions for improving the book will be gratefully received and acknowledged.

Author

Contents

1

Introduction of Plant Pathology

DEFINITION OF IN PLANT DISEASE

In nature for any variety or species of plant there is a set of environmental conditions which favours its optimum development. In nature, however, such an ideal situation seldom exists, and in consequence every variety of plant is subjected to the vicissitudes of the environment. When one or more factors in the environment become particularly unfavourable, the development of the plant is an in some way, and in comparison we growing in the usual range of environment, it a 2 Pears to have abnormal characteristics. Diseased plants are distinguished by changes in their structure or physiology of local price which are brought about by unfavourable environment or by one or another parasitic-agency.

There is no clearly defined line of distinction between normal or healthy plants I and abnormal or diseased plants. Plant disease, therefore, is, like many biological phenomena, difficult to circumscribe or define. We may, however, think of diseased plants as those which have become altered in their physiological and their morphological development to such a degree that signs of such effects are obvious. These external signs on the plant which are characteristic of a given disease are known as symptoms. Whether all deviations from the normal in structure and/or function constitute disease is largely a matter of definition and depends in any case upon what one chooses to dieting as normal for the particular variety or species in question.

Recognition of abnormalities in plants goes back to the early historical records. In general they were attributed to the acts of supernatural beings, to vicissitudes of weather, to insects, or to parasitic seed plants. Only since the middle of the nineteenth century have parasitic microorganisms been accepted as causal agencies in plant disease. In the last two decades of the nineteenth century viruses were recognized as infectious causal agencies, but only now has considerable evidence been brought forward to show that viruses are nonliving but infectious proteins.

Since we now recognize that disease in plants may be brought about by unfavourable environment, or by infectious agencies which mayor may not be living, it is well that we define at the outset certain terms which will be used repeatedly in this book.

COMMON DEFINITIONS

First of all it is well for the student to have a clear concept as to what constitutes the *cause of disease*. This is particularly true because the term is used rather loosely in much of the literature on plant pathology. It is relatively simple to conceive of one or more, or a combination of several, environmental factors becoming the cause of a disease when they alter the physiological processes and morphological development of the plant concerned. When a parasitic microorganism enters into the complex, as is the case in a large percentage of plant disease, there is a common tendency to refer to the microorganism as the-cause of the disease.

Since we now know that variation in the environment is sometimes necessary to predispose the plant, i.e., cause it to become susceptible to the microorganism, one can readily see that the microorganism is not the sole causal factor. Moreover We also recognize that the microorganism, also entity, is subject to the influence of environment, and it often, if not always, requires a certain range of favourable environment to produce infective organs and for such organs to infect the plant. After a microorganism has successfully infected a susceptible plant, the subsequent interaction of plant and microorganism which results in a disease of the former is also environmental influences which determine whether disease develops at all and if so, at what rate and to what degree. It is evident that in the case of a disease in which a bacterium or fungus is concerned there are various other factors besides the latter which enter into the causal complex.

Strictly speaking, therefore, it is not correct to refer to a given microorganism as the cause of a given disease, since this implies that it is the sole cause. However, such usage is common in current literature on the subject. It has been suggested that the organism be referred to as the *incitant,* implying that it incites the disease under the influence of other factors. The term *causal organism* is also proper since it implies that the organism is a part of the causal complex.

Most causal organisms are parasites. In current usage a *parasite,* as applied to an organism, is one which subsists in whole or in part upon living tissue. Conversely, a *saprophyte* is an organism which subsists upon dead organic matter and inorganic materials. A pathogen is an agency which incites disease.

While the term might be applied to any or all of the causal factors described above, it is practically always used to denote the living entity in the causal complex. The terms pathogen and parasite are not synonymous, however. In some cases organisms are parasitic without becoming causal factors in disease.

In other cases the by-products of a Retract saprophyte are an important part of the causal complex, and thus a sarrophyte may Indirectly become a pathogen. *Pathogenicity* is the property of a microorganism whereby it may become a part of the causal complex. *Pathogenesis* is the process or chain of events whereby disease development takes place.

A not uncommon habit among plant pathologists is to refer to the disease concerned by the scientific name of the causal organism. This is, of course, strictly incorrect. The student of palp to pathology should at the beginning develope a habit of referring to the disease and to the causal organism as distinct entities.

Diseases of econorllic plants have usually acquired one or more *common names*. As a rule such names are descriptive of some important phase of the disease, e.g., late blight of potato wilt of tomato; black stem rust of wheat; club root of cabbage. Sometimes the generic name of the causal organism is used as a descriptive adjective in the common name, either because a sufficiently descriptive common word is not available or because it is necessary to distinguish between similar appearing diseases with different causal organisms. Thus wilt of tomato may be the result of the action of distinct causal organisms; hence, Fusarium wilt of tomato, Verticillium wilt of tomato, bacterial wilt of tomato.

The Rhizoctonia disease of potato is another example of a case where the genus of the parasitic phase of the causal organism is used in the common name of the disease. This organism has a wide range of pathogenetic effects upon the potato plant. Under certain conditions basal stem cankers are formed, the result of which may be upward rolling of leaves, abnormal purpling of the foliage, deposition of starch in axillary buds of the stem to form aerial tubers.

Quite apart from the disease on aerial parts of the plant, there are formed superficial black resting bodies on the surface of the potato tuber. Because the last symptom is a very obvious one, the disease is often referred to as black scurf. However, since- the last term describes only one phase of a wide range of distinct disease reactions, the first termentioned, Rhizoctonia disease, is also used.

The causal organism, if fully described, has been given a specific name within a given genus according to the accepted system of plant nomenclature. Since the generic name and specific name are Latin words always printed in italics, they should always be underlined in any handwritten or typed manuscript. This name of the organism is referred to as the Latin binomial of the organism.

The species as originally described may be later transferred to another genus by the same or another investigator. In such a case the species name is retained unless it has been declared invalid for some legitimate reason. The name, or an abbreviation thereof, of the investigator who first described the species follows the binomial. If the species is referred by another investigator to another genus, unless the transfer is from a form genus to that of a perfect form, the name or abbreviation of the one making the original description is retained in parentheses immediately after the binomial, while the name or abbreviation of the one responsible for the current binomial follows.

Since there is not always agreement as to the proper genus to which a species belongs or as to whom the first authentic description of a species is properly attributed, there may be found more than one binomial for a given organism in current literature. On this account it is customary in a paper dealing with a given disease to give the complete binomial with the proper authority when the organism is first mentioned. This indicates clearly to the reader which authority the writer chooses to follow. Subsequently in the paper the authority may be omitted, and some writers prefer to make subsequent reference to the causal organism by using the common name of the disease as a descriptive phrase modifying "organism," "fungus," "bacterium," etc., e.g., the crown-gall organism, the potato late-blight organism, the potato-late-blight fungus, the fire-blight bacterium.

In certain cases usage has resulted in exceptions to the general rule, in that the same word is applied to the disease and to the causal organism. Most common examples are found in the rusts. The large group of diseases known as rusts acquired this name long before organisms were recognized as causal agencies, and the term came into use because it described a prominent characteristic of the diseases as a group. Like wise the organisms were known, and many were described and named, long before they were recognized as pathogenic agencies. Because of their characteristic colour and the consistency of the spore mass, this group of organisms was also referred to as rusts. Thus the same word came into use for disease and for pathogen long before their relations were fully understood. The same chronological history is true for the smuts. It is the more important, therefore, to connote whether the term is being used to apply to the disease or to the causal organism.

In the case of diseases in which parasitic organisms are a part of the causal complex the diseased plant is commonly referred to as the host. While usage has made this the most widely used term, objection has been raised that it does not imply the actual relation which ordinarily exists between the plant and the pathogen. The term suspect is a newer word coined to apply to the plant which is subject to a given disease brought about by a given causal complex.

The above terminology does not always strictly apply in the case of virus diseases. While the inanimate nature of viruses is not accepted in all quarters and practically nothing is known about their mode of increase or reproduction, they have above all in common with pathogenic microorganisms the property of being transmitted from plant to plant to become a part of the causal complex. While several schemes of nomenclature for plant viruses have been proposed, none has been fully accepted as yet. Therefore in current literature the student will find a great diversity among authors in method of reference to a given virus. They are not universal referred to as pathogens, and their disease-producing property is not commonly referred to as pathogenicity, largely because of the doubt that they are living entities.

One system known as the Holmes system sets up a scheme of genera and species whereby each described virus is given a binomial. While this system has many admirable features, it is nonconformist in that it uses system set up for living organisms to classify what may well be nonliving organic compounds, which cannot be defined on the bases used in the binomial system for living organisms. An earlier suggestion by Johnson is to give each virus described from a given suspect or host consecutive numbers, e.g., potato virus 1, potato virus 2, cucumber virus 1, etc. A modification of the Johnson system, proposed by Smith, substitutes the generic name of the host for the common English name, e.g., Solanum virus 1, Solanum virus 2, Cucumis virus 1, etc. Some writers follow one, some another, of these systems, while still others avoid all I three systems and refer to the virus concerned by means of the disease with which it is associated, e.g., common-tobacco-mosaic virus, potato ring-spot virus, tobacco-ring-spot virus, spotted-wilt virus, potato-witches broom virus, etc.

When a pathogen is a part of the causal complex, the organism is first present in the environment in some overwintering form. The overwintering stage may be in or on seed of the host, in perennial wild hosts, in debris of infected plants of the previous season, as dormant spores, sclerotia, or other organs in the soil, in the form of an inhabitant of the soil flora, and in various other forms. Whatever this source, it becomes the *source of primary inoculum.* It may serve as the infective agent and thus become the*primary inoculum,* or it may multiply as a saprophyte to produce spores which become the primary inoculum. When the host becomes diseased, the organism commonly produces another crop of spores or infective bodies, which are known collectively as the *secondary inoculum* and serve to cause *secondary infection*. This last cycle may then be repeated several times during the current season.

The first step in the relation of the inoculum to the host consists of penetration. This may take place in one of a number of ways which are characteristic of the organism and are influenced sometimes by the environment. The various methods- of penetration will be discussed in later chapters. It is important to recognize here that there is a distinction between penetration and infection *Penetration* refers to initial invasion of the host by an organism. We know that many organisms penetrate cells of plants which are not susceptible and which do not respond with any sign of disease; moreover they die without the stage of penetration. Thus we may have many cases of penetration without infection. *Infection* implies the establishment of the pathogen within the host following penetration. It does not imply the production of disease symptoms. There may be a long period between infection and appearance of disease, but ordinarily one follows rather closely upon the other. The interval is known as the *incubation period*. Just as penetration and infection are to be regarded as distinct, so infection and disease development are distinct. *Disease development* refers to the events which occur between the time of infection and the complete expression of disease. In general, all disease development is a protracted process which involves production of secondary inoculum and progression of external evidences of disturbance of the morphological and physiological processes of the host plant; these external evidences are referred to as *symptoms.* The chain of events in disease development is referred to as the *disease cycle*. While, when a pathogen is involved, the disease cycle is intimately associated with the organism, it is distinct from the *life cycle of the organism.* In the case of many causal organisms only part of the life cycle is passed in association with the host, the remainder being in association with dead host tissue, independently as a saprophyte, in association with an alternate host, or in association with an insect vector.

An *infectious organism* is one which may be transmitted from a diseased host plant to a healthy plant and which is capable of inciting the disease under favourable environment. An *infectious disease* is one in which a transmissible causal organism is a part of the causal complex Virus diseases are also usually referred to as infectious since the incitant agency can be transmitted from a diseased to a healthy plant and.initiate the disease in the latter. In the light of present knowledge it is not proper to group them with infectious organisms, since there is no evidence that they reproduce themselves, whereas it is entirely possible if not the infective entity acts as a catalyst which stimulates the formation of virus protein instead of normal host protein during the period of development.

Various *levels of parasitism* among microorganisms have long beer recognized. In many cases two living organisms may live in association with one another, and in a sense each may be parasitic upon the other However, they are not only nonpathogenic to one another but also are each essential to the other's development or at least each is beneficial to the other's development. This phenomenon is known as *symbiosis* and the participating organisms are known as *symbionts*. Lichens are classic examples of symbiosis a fungus and an alga living in close association with one annotate. The alga thoughts photosynthetic property converts kinetic energy of the sun's rays into potential energy of carbohydrates useful to its own development and essential to the life of the fungus. The latter through its property of producing hydrolytic enzymes breaks down materials in the substrate, making energy available and providing mineral and organic nutrients essential to itself and the alga. Another case of symbiosis is that of the nodule-forming bacteria in relation to their respective leguminous hosts. Here the bacterium penetrates and infects the root hair; the host responds by abnormal proliferation of cells to cause a gall, or to this point the process might be considered pathogenic.

However, the bacteria are capable of fixing the free nitrogen of the air and converting it into the essential nitrogenous compounds in their cells. Presumably, as the bacterial cells disintegrate, the nitrogenous compounds become available to the legume plant. Since higher plants are generally incapable of utilizing free nitrogen, an element essential to protein formation, this symbiotic phenomenon is a great economic source of available nitrogen.

That symbiosis is a phase of parasitism and is very close to pathogenesis is shown by further study of the legume-bacteria phenomenon. All strains of the bacterium adapted to a given legume host are not equally efficient in nitrogen assimilation. In fact, some strains show little or no evidence of such a property. When such a strain is allowed to infect its legume host, nodules develop as usual, but in the absence of nitrogen assimilation thc usual benefits to the host are lacking, and the latter shows certain disease symptoms when compared with uninucleate plants. Thus the legume bacterium is in this case not symbiont but a pathogen, the parasitism involved results in pythonesses rather than symbiosis.

Contiguous to parasitism but distinct from it is the phenomenon referred to here in as *eplphytlsm*. An *epiphyte* is a plant growing upon another plant, and, as the term is usually applied, it refers to cases where no peridotic or truly symbiotic relation is involved. Particularly in tropical religions, many higher plants grow on shrubs or trees, not in association it the soil. They secure essential mineral nutrients from the dead bark issue of the substrate plant but do not penetrate the living tissue of the apparently, have no beneficial or detrimentaleffect upon the Oscillator. While some parasitic microorganismsare largely epiphytic, they are not considered as members of this group. Opposed to epiphytes are *endophytes,* which are contained entirely within the substrate plant.

Having delineated symbiosis and epiphytism in relation to *parasitism,* we may now consider various grades of the last. After the relation of microorganisms to plant disease became fully recognized shortly after the middle of the nineteenth century, it soon became-apparent that pathogenic organisms vary widely in the manner in which they bring about pathogenesis under the influence of favourable environment. Parasitism is a very complex phenomenon which is basic to the understanding of plant disease.

Its various phases will be referred to repeatedly in subsequent chapters. Here it is essential only to explain the important terms used throughout the book. Some pathogens, so far as is shown, cannot complete their life cycles except in a parasitic relation to the respectively.

They may persist as dormant spores away from the host and such organs may germinate and produce secondary spores away from the host.

Thus part of the life cycle may proceed apart from the parasitic relation. Infection must occur, however and parasitism obtain for the completion of the life cycle. Such organisms are known as *obligate parasites,* and they are Bound in various groups of the fungi one are known among the bacteria.

Viruses have properties in common with obligately parasitic organisms in that they increase in amount only in association with living suscepts. Common examples of obligate parasites are the rust fungi the powdery-mildew fungi, the downy-mildew fungi, the crucifer-clubroot organism, the potato-black-wart fungus. *Obligate saprophytes* are those microorganisms which have no relation with living cells and secure their nutrients from dead organic tissue or from available inorganic materials. A large share of the pathogens, however, have parasitic and saprophytic phases of their life cycles. The fact that an organism is in part a saprophyte does not in any sense detract, necessarily, from its potentiality as a pathogen. Many organisms can live for indefinite periods as saprophytes in pure culture or otherwise and still remain highly virulent pathogens.

There are, however, many organisms which are in the main saprophytes but have weakly pathogenic properties which an ordinarily expressed only when particularly unfavourable conditions have predisposed the host in such a way that it is unusually susceptible and when such environment is particularly favourable to the expression of the weakly pathogenic properties of the organism. Such organisms are called *facultative parasites*, which term implies that they are ordinarily saprophytic but under certain conditions may become parasitic. Many damping-off organisms are typical facultative parasites.

At the other end of the wide range of parasites that are not obligate parasites in the strictest sense are the *facultative saprophytes*. These organisms ordinarily have life cycles similar to the obligate parasites, in that they pass most of the cycle in association with the host and do not subsist as saprophytes. They may, however, subsist as saprophytes for part of their life cycle but do not complete their life cycle away from the host. Many of the smut fungi answer the requirement of this definition of a facultative saprophyte. It should be borne in mind, however that there are no well-defined limits between facultative saprophytes and facultative parasites. It should be emphasized that they represent the extremes in the range of microorganisms which may have parasitic and saprophytic phases in their life cycles and particularly that, as defined, the facultative saprophyte has a much higher potentiality as a pathogen than does the facultative parasite.

Predisposition may be defined as the effect of one or more environmental factors which makes.a plant vulnerable to attack by a pathogen. It is a process which antedates penetration and infection. *Susceptibility* refers to a condition of a plant in which it is normally subject to attack by a given pathogen. The susceptibility of a given individual, variety, or species may be increased or decreased by environmental factors. It may also be higher or lower, as between individuals, varieties, or species because of differences in the inherited characteristics which affect susceptibility. *Resistance* may be regarded as the counterpart of susceptibility and also may be influenced in its expression by environmental factors.

BASIC CLASSIFICATIONS OF PLANT DISEASES

Plant diseases may be grouped in various ways. The method of classification depends to a considerable extent 'upon the purpose of the treatment. In the chapter on History of. Plant Pathology it will be seen that before the acceptance of microorganisms as "causative agencies in plant disease many writers speculated more or less on possible causal factors, but they based their schemes of classification upon effect rather than upon cause.

After 1850 books upon plant diseases grouped them according to causal factors rather than according to effects. Classification on the basis of cause has many advantages in generally text, one of the chief of which is that the most satisfactory approach to a basic understanding of disease is through an unraveling of the various causal factors and their interrelations, as well as their interaction,with the host. For this reason such a scheme is used in this book.

In a manual of diseases of a certain group of plants, the consideration of classification of diseases according to the host plant has decided advantages, and such a scheme is often used, e.g., cereal diseases, truckeprop diseases, corn diseases, flax diseases, etc. In other treatises diseases advantageously treated according to symptoms or effects, e.g., root powdery mildews, wilts, leaf blights, overgrowth, or hyperplastic diseases, hypoplastic diseases, necrotic diseases, etc.

Parasitic and virus diseases are often classified, in occurrence, into three groups: endemic, epidemic, or epiphytotic, and sporadic.

ENDEMIC DISEASES

When a disease is more or less constantly present from year to year in a moderate to severe form, it is classed. This is usually based on the fact that the pathogen or virus is well established and long persistent through its ability to survive from one crop,season to the next in soil, on crop 'refuse, or in wild hosts.

It is also implied that the environmental conditions are usually favourable for inoculum to develop infection to take place, arid disease development to ensue. Diseases due to pathogens which persist for long periods in the soil, such as those of cabbage yellows and onion smut, are commonly endemic. When the infectious agent readily survive and in prevarlent wild hosts, as is the case with cucumber nomic in milkweed or pokeweed plants, the disease is also usually endemic.

EPIDEMIC, OR EPIPHYTOTIC, DISEASES

An epidemic disease is one which commonly occurs widely but periodically. The term "epidemic" is derived from a Greek word meaning "among the people" and in the strictest sense applies to diseases of man. The term "epiphytotic" has been coined to' carry the same connotation with reference to plants. Epidemic diseases are often very responsive to variations in environment. The pathogen may be more or less constantly present, as in endemic diseases, but the environment favourable to development occurs only periodically.

It is thus evident that the environment in such a case may well be the major determining causal factor. Epidemic diseases "may also occur where the environment is usually favourable but where the pathogen is irregular in its occurrence. This may be because of the inability of the organism to survive always from one crop to the next. It 4 may be dependent upon seed or other propagative parts of the host for introduction. Insect vectors may be necessary for introduction or spread Epidemics of black rot of crucifers are commonly dependent upon introduction of the pathogen with the seed. Late blight of potato is common epidemic in nature because of its sensitivity to fluctuation in temperature and humidity. Curly top, a virus disease of sugar beet in the Rock Mountain states, is dependent upon the introduction of the virus by hoppers which overwinter in more or less distant areas. If the winte reduces greatly the overwintering insect population, infection and spread are minor; if the conditions are the reverse, early and continued infection results in severe disease development.

SPORADIC DISEASES

Sporadic diseases in reality belong to the epidemil group. The term is applied to those which occur at very irregular interval: and locations and in relatively few instances.

It should be borne in mind that a given disease may be endemic in region and epidemic in another. Late blight of potato is endemic is certain parts of northern Europe because the environment is practical always favourable at some time during the season. It is epidemic in the upper Mississippi valley and on the winter crop in the Gulf states, because some seasons are entirely unfavourable for its initiation and development.

DISEASES CLASSIFIED ACCORDING TO MAJOR CAUSAL AGENTS

There are two methods by which major disease classes may be set up according to cause. By one method, diseases are divided into two classes infectious and noninfectious. In this case all diseases incited by microorganisms, parasitic phanerogams, nematodes, and viruses are regarded as infectious. However, it may be questioned whether certain of the insect-incited diseases are infectious, although they are parasitic. By the second method, diseases are divided into nonparasitic and parasitic classes. In this classification some authorities would place virus diseases in one class, and some would place them in the other.

I. *Nonparasitic diseases*: Diseases incited primarily by,
   1. Low temperatures.
   2. High temperatures.

3. Unfavourable oxygen relations.
4. Unfavourable soil-moisture relations.
5. Injurious atmospheric impurities.
6. Lighting injury.
7. Mineral excesses.
8. Mineral deficiencies.

II. *Parasitic diseases*: Diseases incited by,
1. Bacteria.
2. Plasmodiophorales.
3. Phycomycetes.
4. Fungi Imperfection.
5. Ascomycetes.
6. Basidiomycetes.
7. Phanerogams.
8. Animal parasites.
   a. Nematodes.
   b. Insects.

III. Virus diseases.

2

Plant Diseases Development

The earliest records of man reveal that as he began to depend upon cultivated food, feed, and fibre plants for his livelihood, the problems of eroplosses, food shortages, and famine continually arose to plague him. Greek philos la, oppers, such as Theophrastus noted the occurrence of crop maladies and speculated as to their cause and cure. Down through the Middle Ages we find that scholars were impressed by the appearance II of plant diseases but were quite confused as to the factors which brought them about. Scientific experiments were not common, and the occasional keen observer of nature was hampered by the current custom of explaining natural phenomena by the process of deductive reasoning.

Incorrect interpretations, based on unsound analysis of facts and influenced by superstition and religious dogmas, were passed from one generation to another until they became accepted as hard and fast laws of nature. At the close of the eighteenth century, about 150 years ago, there was no evidence that our modern concepts of the nature of disease inception in plants and the relation of environment to disease development had been I adopted by those concerned with plant science and with plant culture.

DEVELOPMENT OF MYCOLOGY

Modern plant pathology did not develop alone; it grew along with and was influenced by related sciences. One of the most important of the allied fields was mycology. The fleshy fungi were,thought by the Greeks and Romans to arise de novo from dead organic matter under the influence of the elements. This idea" came naturally since all the early philosophers believed firmly in the idea of spontaneous generation of lower animals and plants. We shall find that the tenacity with which this theory was I held until the middle of the nineteenth century was a dominant factor in the delay of the development of our present concepts of parasitism and disease.

In the sixteenth, seventeenth, and eighteenth centuries some prominent it botanists were skeptical of spontaneous generation. Among these were Malpighi, the pioneer in plant anatomy, and Jung, pioneer plant systematists. On the other side were Cesalpino, another pioneer plant systematist, and Hooke, an early microscopist. The discovery of bacteria by the Dutch draper, Leeuwenhoek, seems not to have influenced current thought on reproduction of lower organisms.

A contribution which stands out in this period, is the work of Micheli. This Italian botanist studied both the higher and lower forms of fungi. In his "Nova plantarum genera" he described many new genera and illustrated their reproductive structures. He was the fist to collect spores of fungi, sow them on organic substrate and watch the development of mycelium together with sporangia and spores characteristic of the species. He was convinced that fungi arose from their own spores. However, even these careful observations and the conclusions inductively arrived at were not accepted generally.

In 1748, Needham, an English clergyman and naturalist, upheld strongly the theory of spontaneous generation. To prove his point, he boiled meat in flasks, corked them, and set the containers away for a timc, only to find that the contents had spoiled due to the microorganisms. In 1775, Spallanzani, an Italian investigator, challenged Needham and set out to prove that decay meat by bacteria and other lower forms of life could be prevented by heating tnematerial in a flask which was sealed off in such a way as to exclude contamination from the air. His experiments supported this contention, but others failed to secure similar results consistently. These failures, we can now understand, were due to what would presently be regarded as faulty technique. This pioneer investigator likewise did not receive recognition and was successfully refuted for the time by his opponents.

The predominant attention of botanists from the time of Cesalpino was directed towards naming and describing plants rather than searching into the phenomena whereby- they grew and developed. By the turn of the eighteenth century the dogma of the constancy of species had become deeply ingrained in the minds of men, and this concept was fostered by the ecclesiasts, who found it to cosicide well with the prevalent religious teachings. For 150 years or more, men had struggled to find a rational basis of naming and classifying plants. Linnaeus was the leader in the eighteenth century who established the presently used Latin binomial system, his work culminating in the publication of "Species plantarum" in 1753. This Swedish botanist had a wide influence. His students and followers were collectors and classifiers, but they were not experimenters. Micheli had been interested in morphology and taxonomy of the fungi; but he had also given attention to their physiology and reproduction. He was ahead of his time. Linnaeus was a product of scholasticism and a firm believer in the dogma of constancy of species. He assented to the view that fungi arose through spontaneous generation.

Persoon worked chiefly in France; Schweinitz in eastern United States; Fries in Sweden; Nees 4:9 von Esenbeck in Germany; Leveille in France. Persoon's "Synopsis methodica fungorum", pub1ccu: and lished in 1801, and Fries's "System a mycologicum", published in the opr. years 1821 to 1832, became the foundation stones of the modern application of the Linnaean system to the taxonomy of the fungi. In the main, the these men were concerned with the naming of fungi and with their class and they were not concerned with their reproduction and with the their relation to plant disease. Although Persoon gave the name Puci ke Ecinia graminis to the fungus associated with the black stem rust of wheat and, in 1794, he did not regard the fungus as the incitant of the disease. In 1818 he expressed the opinion that some fungi grew from spores, but that SU others arose by spontaneous generation.

In 1807, Prevost observed germination of the spores of the wheat bunt organism and conceived the idea that this organism penetrate the young wheat plant and was the actual cause of the disease. His revolutionary idea was refuted and did not gain favourable attention until 40 ad years later. Although predominant attention was given by most my cologists to taxonomy, there began about 1820 a renewed interest in the type of study that Micheli had undertaken 100 years earlier.

Ehrenberg brought together known information on propagation of fungi and attributed further observations of his own on spore germination, as well as al the first description of sexuality in fungi. Nees von Esenbeck repeared al Micheli's experiment by sowing bread-mold spores on bread and later obtaining ripe sporangia. Dutrochet connected the sporophores of fleshy fungi with the subterranean mycelial strands which had been a considered up to that time as distinct fungi.

PREMODERN HISTORY

During the seventeenth century, plant diseases were increasing in importance in Europe with increase in-population and in intensity of plant culture. Farmers, by and large, accepted them as inevitable concomitants of unfavourable soil and climate. Some growers, however, tried to do something to alleviate the damage, for we know that treatment of wheat seed with salt brine to control bunt began in that century. The association of the barberry with stem rust of grain had, been recognized by farmers for centuries. It has been claimed by some that in 1660 farmers in the vicinity of Rouen, France, secured passage of a law which required destruction of the barberry bushes in that area.

The authenticity of this report has never been firmly established. We are certain, however, that colonists took the barberry to New England with them and that wheat rust became increasingly destructive after 1660. A law was passed in Connecticut in 1726 which authorized town meetings to adopt barberrye radication measures. Similar legislation was enacted in Massachusetts in 1755 and in Rhode Island in 1766. It should be emphasized that in the development of the control practices mentioned above there is no evidence that anyone had the present-day concept that many diseases are brought about by separate infective entities under the influence of favourable environment.

Theophrastus recorded his observations on the effects of various diseases of plants and in the main assumed that the basic causes lay in the vicissitudes of the weather. Pliny the Elder wrote upon the diseases of trees and of the cereal grains, but his work is little more than a translation of that of Theophrastus. Giambattista della Porta's discussion of diseases of plants, consisted largely of quotations from Theophrastus and Pliy, with some observations of his own. In 1705, the" French systematist Tournefort classified plant diseases, dividing them into two major groups, one due to external causes, the other to internal causes.

There followed a period of 150 years in which most writings on plant diseases consisted in descriptions of their appearances or symptoms and an arrangement of them into classes, genera, and species. The influence of the Linnaean binomial system was- evident in each of these compilations. Adanson in his work on "Familles des plantes" enumerated 23 species of plant disease, which were divided into two classes unpending upon whether they were due to internal or external causes. The nature of the causes was however, primarily a matter of speculation.

In Rozier's *"Cour complet dagriculture"* the list was expanded to 2 32 species. Plenck divided diseases into eight classes: external injuries, flux of juices, debility, cachexias, putrefactions, excrescences, monstrosities, I sterilities. In 1773, Zallinger placed plant diseases in five classes: phlegmasiae, paralyses or debilities, "cudischarges and drainings, cachexia, or bad constitution of the body, chief defects of different organs. Fabricius, a Danish botanist who student. Wied under Linnaeusand was for a time on the staff of the Royal 'Bot anical Garden in Copenhagen, published a classification of plant diseases theses in 1774. He spent much time in Paris, and evidently drew much of an his material from Adanson's book. His treatise is an arrangement of diseases into classes, genera, and species on the basis of their symptoms. 1: of That Fabriciu as not satisfied with current explanations of certain plant diseases is indicated by his statement that "certain It is that the of cases and symptoms of smut can never 11 better explained than by as summing something organized to be the cause.

In 1755, Tillet director of the mint at Troyes, France, and amateur botanist and agriculturalist, published an important paper If on wheat bunt. He found that opinion varied widely as to the cause of 1 the-disease-evil fogs, winds, excessive soil moisture; improper plant nutrients, soil texture, insects. In well-planned field experiments he tested the effect of dusting seed with the black mass contained in diseased kernels, and of treating such seed' with water containing- common salt and Plime. He demonstrated that the disease was contagious, since dusted a seed produced a much larger percentage of diseased plants at maturity than did undusted seed, while seed treated with salt and lime produced a lower percentage of diseased plants.

Tilet did not recognize that the smut dust consisted of spores which were "seeds" of a parasitic fungus. He rather looked upon the spore dust as merely the carrier of a poisonous entity. However, he did establish for the first time that the disease was 2 in some way contagious through the smut dust and that healthy seed produced only healthy plants. Furthermore, he demonstrated clearly that chemical treatment inhibited the contagious entity. In 1759, Count Francesco Ginanni in Italy published a book on the same subject entitled *"Trattato delle malatie del grano in erbo"* in which he attributed the cause of bunt to minute Insects. Another Frenchman, Tessler, working in the same period, published a paper entitled "Traite des maladies, It des' grains," in which he attributed rust to a stoppage of the transpiration 'I of the plant, caused by mists.

In 1766 an Italian, Giovanni Targioni Tozzetti, put forward al theory that the rusts and smuts of wheat consisted' of the vegetation of fungi arising under the epidermis of the plant. Fontana an Italian physicist, published a paper the following year in which he expressed the view that the grain rust was a distinct parasitic entity.

This new theory received some support, no doubt, but the weight of opinion was against it. In this same decade Anderson, an Englishman, wrote concerning the degeneration of potato varieties which we now know to be due to infectious viruses. In 1778 he published evidence to show that the malady was tuber-borne and likened it to smallpox in its infectiousness.

Zallinger in his book on plant diseases, "Demorbis plantarum" upheld the prevalent concept that fungi found in association with plant-disease lesions were abnormal structures of the diseased plants.

MODERN PLANT PATHOLOGY

It is apparent that at the beginning of the nineteenth century when Persoon and, a little later, Fries were actively naming and compiling fungi without reference to their relation to plant disease, those who were writing about these maladies were still not cognizant of the true relation to them of microorganisms. The first investigator to bring forth the facts which clearly demonstrated the pathogenic nature of any microorganism was Prevost, a Swiss professor of philosophy at the Academy at Montauban, France. He studied the wheat-bunt disease Jor about 10 years beginning about 1797. He published in 1807 his epochal "Memoir on the immediate Cause of Bunt or Smut of Wheat, and of Several Other Diseases of Plants, and on Preventives of Bunt". After presenting an accurate description of the disease, he reported important results of experiments upon it.

He studied and described the germination of the spores. He demonstrated, in confirmation of Tillet, that spores applied to the seed in some way resulted in infecting the plant. He studied the growth of the- fungus within the growing wheat plant and its final development in the embryo where spores were formed, and where the life cycle of the fungus was completed. He showed that solutions containing copper sulfate prevented spore germination. He distinguished between fungicidal and fungistatic effects. From these facts he demonstrated the basis of control through seed treatment, a practice which had been carried on empirically and without any rational explanation for 100 years and more.

The thorough hold of the theory of spontaneous generation on men's minds at this time was demonstrated by the fact that Prevost's results and interpretations were not accepted. A committee of the French Academy rejected his conclusions as unsound, and some 40 years had to elapse before others confirmed Prevost and pressed the adoption of the germ theory.

"In the period from 1807 to 1845 the prevalent teachings are illustrated in the writings of Count: Fillipo Re, Franz Unger and Franz-Julius Ferdinand Meyen. Re was professor of botany and agriculture at the University of Modena, Italy. His treatise on the diseases of plants was published in 1807, in the same year as the publication of Prevost's memoir. A second edition was published in 817, and, in 1849 Berkeley published an English translation of the latter in the Ci Gardener 8 Chronicle, stating that it was the best book available on the subject. It is of interest to take more than casual note of Re's book, because it portrays the increasing conflict of thought on the nature of plant disease in the beginning of the nineteenth century. Diseases were divided into classes and genera on the basis of their symptoms and their a supposed causal factors.

The plan was not fundamentally different C from that of classifications Tesented in the last half of the previous century. The influence of current writings on the classification of human, diseases is evident. We need mention here only the titles of the five classes: diseases constantly sthenic; diseases constantly asthenic; diseases derived, sometimes from the sthenic, sometimes from the asthenic state; mechanical injuries, wounds; indeterminate diseases. In the last class were those diseases which author: regarded as of uncertain origin.

The pressure of the parasitic theory is evident from the remarks of Re who evidently did not see the work of Prevost before He second edition of his book was Published in 1817. The experimental evidence was racking, except that presented by Prevost, to clinch the germ theory. The adherents of the autogenetic theory of disease still had the better of the argument.

Franz Unger was born in Germany but spent most of his life in Austria Hungary. He was trained in medicine but in early mid-life turned to botany, becoming a professor of botany in Graz in 1835 and. professor of vegetable physiology in Vienna in 1849. In 1833, while serving as a medical officer in the Tirol, he published a book entitled "Die Exantheme 2 der Pflanzen". He believed that the fungus organisms associated with disease were endophytes but not parasites, that, they had no independent existence, and that under certain environmental conditions every plant had the potentiality of producing them, in some way, from the waste material or exanthema of the plant. The influence of human pathology upon the interpretation of plant disease was still strong.

Meyen, a professor of botany in Berlin, was concerned primarily with It' plant anatomy and physiology, but he wrote a look in plant pathology I which was published in 1841. In the year following his death. He followed Tournefort's system of classification of plant diseases. He considered smuts of cereals to be non-infections and to be due to a rotting of to sap brought about by Improper fertilization of the oils Rusts and mildews 4, were likewise regarded as the by-products of disease due to environmental factors. He described the formation of the pustules of white rust o shepherd's-purse in a unique manner.

He claimed that as the disease attacked the plant, cells of the latter enlarged and stretched out in some way towards the epidermis and cut off several spores in succession. Thus this incomplete picture of the development of sporangia as we now know it was misinterpreted to strengthen the autogenetic theory of disease.

About 1830, a few decades after the culture of potato as a food crop had expanded rapidly in Europe and North America, the late-blight disease appeared in both continents.

It reached epidemic proportions by 1845. In that year it received much attention by writers in agricultural papers in Europe and the United States. The editorials and comments by contributors in the Garaener's Chronicle of 1845 are particularly enlightening to the student of plant pathology because they reveal better than any other existing evidence the trend of thought of laymen and scientists on the nature and cause of this destructive disease. While the association of the causal fungus with the blight of foliage and the rot of tubers was recognized by all who studied the disease closely, most writers held, in., accord with the autogenetic theory of disease upheld by Unger and by Meyin, that the fungus was the result rather than the cause of the disease.

Von Martius, in Germany, is one of the first to distinguish the fungus. 'He ascribed the disease to the effects of the fungus and published drawings of the latter. In 1845, Morren, director of the school of agriculture at Liege, Belgium, where the disease had dovastated the potato fields, described the disease and the associated with it He conducted crude inoculation experiments with the fungus and concluded that it was the major cause. He suggested the use of a mixture of lime, salt, and copper sulfate in suspension in water as a drench of the soil to reduce tuber rot.

Had he applied this mixture to the foliage, he might have anticipated the discovery of Bordeaux mixture by 40 years. In the same year, Montagne, a French mycologist, collected the fungus in France, named it Botrytis infestans, and sent specimens to Berkeley in England. The spreading epidemic had not yet reached Berkeley's home in North amptonshire, but it soon did. He confirmed Montagne's diagnosis and description but hedged on dogmatically upholding the idea that the fungus was the primary cause of the- disease.

English horticulturists, led by Lindley, editor of the Gardener's Chronicle, maintained that the unusual cool moist weather had so upset the normal growth of the potato that an internal breakdown of sort followed. The fungus was universally present but entirely secondary, feeding upon and multiplying in the dead plant tissue. Lindley headed a commission to investigate the disease in Ireland, where acute distress and famine followed in the wake of the epidemic. After what Lindley and his associates thought to be a thorough investigation, the weather got the blame, and the fungus remained innocent, in spite of increasing evidence from some quarters, especially in America, that blight and rot occurred even where the weather was not abnormally cool or moist.

In 1846 Berkeley took the stand that Botrytis infestans was the primary cause of the blight and tuber rot of potato. He maintained, further, that wheat bunt was incited by a fungus. In 1848 he reported the potato fungus on Datura stramonium L., on Anthocercis, and on tomato. He stated again with regard to the case of the disease: "I have no more doubt that the malady is caused by the fungus than I have that many of the cereal diseases are caused by fungi."

What was lacking was experimental proof that be infestans was the primary cause of the disease. Morren had provided some evidence, but the techniques of those days were still not adequate to refute completely the interpretations of the autogeneticists. Undoubtedly the supporters of the germ theory were increasing in number. Its general acceptance awaited only more clear-cut experimental evidence.

In the decade of 1840 Louis Rene Tulasne and Gharles Tulasne, botanists at the Museum of Natural History in Paris, began extensive morphological studies of the fungi. Their excellently illustrated-accounts of rusts, smute, and Ascomycetes are classics in mycology. In 1847, they published studies on the wheat-bunt fungus which confirmed the observations of Prevost with reference to spore of germination and relation of the fungus to the' host plant. They extended m their observations to several other smut organisms and to several rust fungi. The work of the Tulasne brothers was followed in 1853 by a comprehensive paper on the fungi, including the smuts and rusts, by Anton De Bary. Trained in medicine, De Bary early: turned to m: botany as a career. This paper, published when he was 22 years of age, up was dedicated to Alexander Braun, his professor of botany at Berlin, and a student of the algae.

He studied the development of several fungi in a critical manner. He Taking first the corn smut, he traced the development of mycelium, sporiferous branches, and chlaydospores. He was impressed by the a; fact that the mycelium in the intercellular spaces was distinct from rather Litoan a part of the host cell wall. In fact he pointed out that Unger, had I published the same observation in 1840. As the spores matured, he, noted that the mycelium passed into the spores and the hyphae becalne indistinct.

He saw that as the cllnsmut gall began to develop, the host cells underwent continuous division up to about the time when spore formation was, initiated. Then cell division slowed down gradually, and eventually) the host tissue was broken down as the rapidly growing fungus drew 1 nourishment' from the cells. Three other smuts were studied, and the c conclusion was drawn that all four species followed the same process in their development and differed only in morphological characteristics. He studied critically the development of pustules of white rusts-of several hosts and again interpreted- the mycelium and sporogenous hyphae as I distinct from the host cell and not arising from the latter as maintained by Meyen.

De Bary examined critically several of the common rust fungi including that of black stem rust' of cereals and that of rust of bean. Here he did not distinguish between summer spores,' and winter spores but described the latter and their formation accurately.

Dimorphism of these species- was not described until the next year by Tulasne. The stages of rusts were till considered as separate After Persoon, few workers opposed the recognition of the smut particles as fungi; however, this did not prevent people from regarding them as products of disease in plants or from attributing the latter to entirely different causes entities by my ologists, and De Bary was in agreement with that concept.

He stated that although the.opinion was universal that barberry was harmful to grain, there was of that the spares.of Aecidium on barberry caused infection of cereals. Although he later was the first to prove this to be the fact, he was quite skeptical of the possibility in. 1853.

Meyen had suggested that the pycnia which produced pycniospores in association with Aecidium, Roestelia, and other aecial rusts were male organs. De Bary discussed this possibility at length but was definite in his contention that no pro of had been offered. He expressed hope that the matter of the relation of pycnia to aecia would be settled promptly since several workers were investigating the subject. As a matter of history, the question was not settled until Craigie presented evidence far heterothallism and explained the function of pycniospores in the rusts in 1927.

Had De Bary publish, only the first two parts of his paper the publication would have stood merely as a substantial contribution to the growing information on the fungi and their life histories and to the tax on omic relations of the farms considered. As to the tax onomy presented in Part II, it is evident that De Bary's attempts at classification were far from final. In 'Part III, however,' he launched into the controversial field of the relation of fungi to plant disease.

Reviewing the opinions which had prevailed from the time of Theophrastus,' he made it clear that the fungi associated with the diseases known as smuts' and rusts were regarded by mast farmers and botanist of De Bary's time as the effects of there, than the causes of disease. 'This, he explained, was because they saw these fungi appear simultaneously with the degeneration or failure of the-hast plants and were unable to study the formation of these degenerative lesions Thus all manner of external factors were' blamed as-causes. As ideas changed other theories arose to explain the association of the organisms with disease. Linnaeus considered the Ostalgia spares as eggs of Infuse oria. Later the similarity of smut and rust particles to fungus spares led him to include them among his fungus genera. Finally Persoon introduced a definite classification in which the genera Aecidittm Uredo, and Puccinia were set up.

In brief outline we have scanned some of the major workers and their contributions in mycology and plant pathology up to the time of De Bary in 1853. We have emphasized that throughout the 18th century and the first half of the nineteenth century the taxonomy of plants was I paramount and the dogma of constancy of species had a strong influence upon the trend of botanical thought. Moreover, the theory of spontaneous generation of microscopic forms and of the fungi generally held sway. Parasitism was not an entirely new concept, for the parasitic phanerogams had been generally accepted as true parasites upon their hosts-before this I date, and De Bary in his first paper referred repeatedly to this fact in his I argument for the acceptance of fungi as plant parasites.

It is important that we realise that other phases of botany also passed although transition stages as did mycology. The approach to natural I science by experimentation and inductive" reasoning was also affecting plant physiology and morphology.

Robert look was a British scientist and philosopher who by 1660 had greatly improved the compound microscope. He noted the cellular structure in charcoal and in cork. He published in 1667, a book under the title "Micrographia or Some Physiological Descriptions of Minute Bodies Made by Magnifying Glasses". Two contemporaries of Hooke were Malpighi, Italian physician and anatomist, and Grew, British physician, who pursued studies of plant anatomy as an avocation. These two men presented results of their studies before the Royal Society of London in 1671, and both continued to publish in later years. Their work was largely exploratory and dealt with grosser features of anatomy. Malpighi publishcd "Λnatomc plantarum" in 1675. Grew published "The Anatomy of Plants" in 1682. The important point of interest about these two men is that they advanced plant anatomy to a point which was not again exceeded for over 100 years.

Anton van Leeuwenhoek was contemporary with Malpighi and Grew. Not highly educated, he assiduously devoted his spare time to the preparation of superior lenses and made available to himself equipment capable of higher magnification than that employed by Ma1pighi and Grew.

Born in Holland, where he carried on his work, he communicated his reports to the Royal Society of London. Besides discovering bacteria he also studied plant anatomy. His work as a whole was not so thorough as the studies of Malpighi and Grew, but with superior equipment he discerned some things which they overlooked, such as the crystals within certain plant cells.

Stephen Hales was the outstanding figure in plant physiology of the eighteenth century. His most quoted contribution is his book entitled "Vegetable Staticks", which was published in 1724. Also in the eighteenth century were Kohlreuter, early plant hybridities, and the Dutch physician, Ingenuous who pioneered in the study of photosynthesis. He was the first to conceive that green plants took up carbon dioxide from the air and gave of oxygen. His work was con temporary with and helped by that of the French chemist, Lavoisier, who discovered oxygen and carbon dioxide.

In the first half of the nineteenth century, De Saussure Swiss physiologist, advanced knowledge of the oxygen-carbon dioxide relation to plant nutrition and of the mineral nutrition of plants, while Dutrochet, also Swiss, studied the phenomenon of miosis. Organic chemistry became a distinct science with the synthesis of urea in 1828. But it should beet in mind that the advances of pioneers- 'just mentioned were not very generally accepted. The pull to revert to deductive reasoning and to glmophical rather than the experimental scientific approach was still strong.

Following De Saussure, Liebig became the dominant leader in plant nutrition and championed the us of mineral fertilizers in plant production. Nageli, German physical chemist and plant morphologist, was one of the great German minds of his period. He pioneered the field of physiology of lower organisms.

During the first half of the nineteenth century there was revival of interest in anatomy. It was in this period that Schleiden and the zoologist Schwann formulated their cell theory. In the field of plant anatomy von Mohl, Meyen, Unger, and Nageli were all active. Robert Browne is generally credited with the first observation of the cell nucleus. I Dujarden, French.zoologist, discovered the material within the cell we now refer to as protoplasm. Purkinje, Bohemian biologist, was; the first to use the term "ptoplasm".

DISINFECTION, FERMENTATION, ANIMAL DISEASE

We have already stated that in so far as lower organisms were concerned, the prey-ailing concepts throughout the eighteenth century were that these organisms arose de novo from inanimate substrate. Needham, a Roman Catholic clergyman in' England, undertook to prove that the theory of spontaneous generation was correct. In 1748 he published an account of his experiments, which consisted of boiling meat in flasks, corking them, and setting them away for a time. The contents eventually spoiled, and Needham, maintaining that all life had been destroyed by the I heat, concluded that the organisms within the flasks arose de novo.

Spall lanzani challenged Needham and explained his results arguing that unheated air had been' allowed to contaminate the flasks. Others maintained that heated air was unfit for microorganisms., Spallanzani answered them by sterilizing air by other moons. Nevertheless his views were not accepted. Apart, in France, stimulated by an offer of 20,000 francs by the French government during the Napoleonic wars for a means of preserving food, worked out a method published in 1810, which I became the basis of the modern canning industry. It heated foods in sealed containers in boiling water, and they remined sterile. However, he did not try to explain the basis of his success. There continued much I speculation on the latter with reference to both food preservation and fermentation. Schulze, Schwann, and Schrfhder and Dusch were workers I who between 1836 and 1854' showed that sterilized vegetable infusions' remained so if the air which passed into the container were first passed I through sulfuric acid, cotton, or a heated tube.

Schwann and Cagniard-Latour discovered the yeast cell and maintained that fermentation was brought about by it.

Berzelius, a Swedish chemist on the other and, in 1839, defended the mechanistic explanation of fermentation, and he was followed in this support b: Liebig, the Germa agricultural chemist. It was not until about 18.6°1 that Pasteur furnished Irrefutable evidence that organisms arise from preexisting living entities and that fermentation is a biological phenomenon rather than a purely chemical one. But it should be borne in mind that Pasteur's conclusions were not generally accepted for many years. Lister, the English surgeon, was, however, much impressed by Pasteur's work and with it as a background pioneered the field of antiseptic surgery.

The germ theory in relation to diseases of man and animals was not yet established. First final proof of the causal relation of the anthrax bacillus to the anthrax disease was published by Robert Koch in Germany in 1876. The conclusions were soon confirmed by Pasteur. In 1882, Koch announced the poured-plate method of isolating bacteria by the use of gelatin as a liquefiable solid medium, which was a great improvement over the dilution culture method used by Passeur. In the same year Koch announced his discovery of the tuberculosis organism.

THE PERIOD OF DE BARY AND KUHN

De Bary started his career as a teacher and an investigator in a destined to bristle with new biological viewpoints, activities, and discoveries. In 1858 Wallace and Darwin presented their ideas based on years of careful observation of variation in plants and animals and proposed their theory of evolution based on natural selection. In 1859 Darwin published his "Origin of Species," and the great struggle between the evolutionists and the adherents of the dogma of the constancy of species had begun. In the same period Such was becoming the outstanding leader in the beginnings of modern plant physiology.

The number of supporters of the germ theory increased markedly after 1850. The Tulasnes were active. In 1853 one of them published a paper on the powdery mildew of grape in which he took a definite stand that the Oidium was a distinct parasite. In 1854 Berkeley began the publication of "Vegetable Pathology," which ran as a serial in Gardener's Chronicle through 1854, 1855, and part of 1856. The majority of this treatise was taken up with accounts of nonparasitic maladies, and the system of classification followed more or less that of Re. However, there was added a class under the name Parasitae, which included the rusts, smuts, mildews, and other diseases regarded as having a fungus as the primary cause. This was the first arrangement of plant diseases in which parasitic diseases were given even foamier place.

De Bary's outstanding leadership drew students from various countries in Europe and America. His laboratory became an active centre of study in the new subject of parasitism. Many of his students became prominent leaders in the field of modern plant pathology. A few will be mentioned here to illustrate the caliber of these men and range of their nationalities. Woron in studied with De Baryat Freiburg in 1859. He became the leading plant pathologist of his period in Russia. Brefeld studied with De Bary in 1872-1873. He became an out standing German leader in pure culture and life histories of fungi. In the field of plant pathology he is best known for his studies on the life histories of the cereal smuts.

Millardet studied with De Bary at Freiburg. While professor of botany at Bordeaux he was a leading student of the diseases of the vine and became most widely known for his discovery of Bordeaux mixture. Farlow studied with De Bary at Freiburg in 1872. He spent his long active career as professor of cryptogamic botany at Harvard University, from which vantage point he exerted an outstanding leadership in the study of parasitic fungi from 1874 until his death. Ward was one of De Bary's last students. He worked with him at Strassburg about 1880. He became the leading plant pathologist of England until his death at the age of 52.

While we cannot take space here to delineate the contributions from De Barry's laboratory, it is important to sketch briefly some of his major investigations, which served to clinch for all time the acceptance of the germ theory. 'Boon after the completion of his first paper he turned to a study of the day-mildew fungi. This became another one earning study of the life cycles and the parasitic nature of this group of fungi. The first two papers were published in 1861 and 1863, the final one in 1881. During this study he became involved with the most notable plant disease- of the time, late blight of potato. The fungus being a close relative of the downy mildews, De Bary was well prepared to place on a sound basis the nature and life history of the organism, which up to that time was in much confused state. He con: frilled the work of Speerschneider. He demonstrated the direct and indirect germination of the summer spores; the infection of tubers by sporangia washed through the soil from the leaves; penetration of the I fungus into the host and its subsequent development after infection; overwintering of the fungus as resting mycelium in the potato tuber. He disproved the claim of others that the fungus had resting spores; he placed the fungus in the new genus Phytophthora. His first paper on late I blight appeared in 1861, his last in 1876.

Another matter still unsettled was the relation of rusts to their hosts, the question of alternate hosts. The barberry had been suspected I as an alternate host of the cereal black-stem-rust fungus for centuries, I but no one had actually brought forth the proof In. 1865 De Bary I announced that he had successfully completed the life cycle of Puccinia i graminis Pers., the black-stem-rust organism by carrying i to the barberry and returning it to wheat. He reported the n ea 1866) that he had found the alternate host of the leaf rust of rye, Prubigo-vera secale. Carleton, and that of the crown rust of oats, *P. coronata*.

In his last major research paper, on the damping-off and vegetable rotting fungus, Sclerotinia sclerotiorum. By., he reported studies upon the physiology of parasitism. This pioneer work was among the first which were concerned with the nature of the interaction of host and parasite. The demonstrated that this fungus had a rotting action brought about by substances which diffused in advance of the myceliium.

De Bary also contributed to the advance of plant pathology through his compilations of current knowledge in the field of mycology. In 1866 he published a book entitled. Morphology and Physiology of Fungi, Lichens and Myxomycetes". This appeared as the second volume of Hofmeister's "Handbook of Physiological Botany." His greatest compilation, "Comparative Morphology and Biology of the Fungi, Mycetozoa and Bacteria," was published in 1884, only 4 years before his death. Translated into English in 1887, it immediately became a valuable reference book in plant pathology.

The thoroughness with which this volume was put together has scarcely been exceeded to the present day. It illustrated the masterut grasp of the subject which De Bary had attained, and in the chapters which are concerned with parasitic fungi it showed how completely the modern concept of parasitism had become established 30 years after he entered the field'. "Lectures on Bacteria," a summery of "the pretenses of knowledge and.opinion concerning the mulch discussed questions; connected with bacteria," was published in 1886. An English translation appeared in 1887, and a second edition thereof was printed in 1898.

Contemporary with De Baryand in many respects an equally important figure in the developed of modem plant pathology was Julius Gotthelf kuhn. Born at Pulsnitz, near Dresden, Germany, his early background was associated with agriculture. His father was a small farmer, and at 16 KUhn became a farm manager, remaining in this profession until he was 30. He studied for a year in the agricultural academy at Poppelsdorf and for a year at Leipzig, where he was granted a doctor's degree in 1856. He lectured during the next year at the agricultural academy at Proskau and then returned to farm management on a large estate in Silesia. Here he published his epochal textbook in 1858, entitled "The Diseases of Cultivated Crops, Their Causes and Their Control". This was the first text to be published in which the fungi were regarded as causal factors in plant diseases. In 1862 Kuhn was called to the Chair in Agriculture at the University of Halle, where he began to build a college of agriculture in a major university. He remained at Halle until his death in 1910.

Kuhn had a great influence on the development and application of control measures. In his textbook he recognized diseases due only to unfavourable environment and those due to parasitic organisms including insects, phanerogamic parasites, and cryptogamic parasites. He gave much attention to the improvement of seed-treatment methods for cereals. He published some 70 papers on plant diseases.

While Prevost had completed most of the steps necessary for modern acceptance of the bunt fungus as the major cause of wheat bunt, he did not determine how the fungus penetrated the host. Neither the Tulasnes nor De Bary added any new information on this subject, although De Bary eventually worked out the penetration of rusts and of the potato late-blight fungus. It was Kuhn who first demonstrated the entrance of the mycelium of the bunt fungus into the wheat seedling and followed its development in the host from that point to the formation of spores in the wheat kernel.

Michael Stephanovitch Woronin was a productive worker in the field of plant pathology from the time of his association with De Bary at Freiburg in 1858-1859. His early papers dealt with chytrids and with sunflower rust. The severity of the club root disease of cabbage in the market gardens of St. Petersburg attracted his attention about 1870. This disease had been troublesome in Europe for centuries, but nothing was known of its cause. Woron in discovered by careful study the unusual type of parasite, new to science, which he placed among the slime molds. Still it was sufficiently different from the saprophytic slime molds to warrant the erection of a new order, family, genus, and species. The organism was named Plasmodiphora brassica. A preliminary report in 1874 was followed by a full report in Russian in 1877, the paper being published in German in 1878. This was the first description of a disease incited by a parasitic organism closely related to the saprophytic-slime molds. In the same paper he described a chytrid, Chytridium brassica, parasitic on the roots of cabbage. Woronin maintained a lifelong interest in the fruit-rotting fungi 'belonging to the genus Sclerotinia and published several papers on them.

Soon after the appearance of Kiihn's textbook, there appeared one by Willkomm in 1866 on diseases of trees and one by Hallier in 1868, entitled "The Diseases of Cultivated Plants". The latter followed the same plan as the book by Kihn. In 1874 appeared a text entitled "Handbook of Plant Diseases" by Sorauer. This book is of special significance since it has passed through several editions up to the present time. In the first edition nearly as many pages were devoted to nonparasitic diseases as to parasitic diseases. In fact, Sorauer looked with concern upon the current trend of emphasizing fungus parasites and their life histories and neglecting the part of the host 'plant and the interaction of host and parasite in development of parasitic disease. He is one of the first to emphasize the predisposition of the host to disease by the influence upon it of preinfection environment.

Oskar Brefeld had early association with De Bary in his mycological studies and conducted much research on the fungi extending well into the twentieth century. He pioneered the development of pure culture technique for the fungi. He published rather full descriptions of his methods in 1875 with additional reports in 1881 and in 1883. His first interests at this time were in studying the complete life cycles of saprophytic fungi. His methods were gradually taken over by others and became a basis for present-day research in plant pathology, along with improvements introduced by Koch, Petri, and others. Brefeld turned his skill later to the culture of the smut fungi, and beginning about I 1880, he continued to study this group for some 30 years. He did more than anyone investigator; work out the.life cycles of the cereal smuts.

Another contemporary of De Bary and Kuhn who contributed greatly to the field of plant pathology was Robert Hartig. He was trained in forestry and was professor of botany at the forestry academy at Eberswalde from 1866 to 1876. He was professor of botany at the forestry school in Aschaffenburg from 1876 to 1878; and in the latter year he transferred to

Munich as professor of botany in the school of forestry and became director of the Royal Forestry Experiment Station. He became the outstanding leader in forest pathology of his generation. He contributed numerous reports on diseases of forest trees and on time decay.,An enthusiastic teacher, he was the author of two important texts in the field, "Important Diseases of Forest Trees" and "Text book' of Tree Diseases".

BACTERIAL DISEASES OF PLANTS

Woronin described bacteria in the root nodules of leguminous plants in 1866. Burrill of the University of Inois was the first to relate a bacterium to the cause of a plant disease. In reports extending from 1878 to 1884 he showed that the fire blight of pear was incited by bacterial organism. It is of interest to note that the first acterial disease of plants was demonstrated only two yeàrs after Koch had finally demonstrated the first bacterial disease of animals. In 1883 Wakker in Holland, published results of a thorough investigation of the yellow slime disease of hyacinth, which was shown to be incited by a bacterial organism.

Arthur, working at the New York Experiment Station, in articles published from 1885 to 1887, confirmed Burrill's work. He isolated the organism in pure culture, using the newly developed plate culture technique of Koch, and fully demonstrated the pathogenicity of the bacterium. Savastano demonstrated in 1887 that a bacterium incited the knot disease of olive. Erwin F. Smith began to study bacterial diseases of plants soon after 1890. His first attention was directed towards the bacterial wilt of cucurbits. His first report was in 1893, and a major I paper was published in 1895. He took up in-succession the brown rot of solanaceous plants and the black rot of crucifers, the latter of which had been first described by Painmel in 1893.

It is of significance that the leadership and most of the-important work on bacterial diseases of plants from 1878 to 1900 and even later was found I in the United States. In fact there was a distinct air of skepticism on I the part of European botanists concerning the importance of bacteria as I plant pathogens. When De Bary published his "Comparative Morphology" in 1884, he was aware of Wakker's work on the hyacinth I disease but apparently did not know of Burrill's work on fire blight. In, the section on bacteria he devoted one short paragraph to those parasitic on plants. He pointed out that they had "scarcely ever been observed.... One reason for this may be that the parts of plants have usually an acid reaction." In "Lectures on Bacteria" of the 146 pages were devoted to bacteria parasitic on plants.

The works of Wakker, Burrill, and Arthur were noted as well as those of Prillieux in France on I the bacterial decay of wheat grains and of Reinke and Berthold in Germany of on bacterial rot of potato tubers. That De Bary still regarded bacteria as of minor importance as pathogens of plants is shown in his statement that "saprophytic bacteria may also, under special conditions, attack the tissues of living plants as facultative parasites, produce disease in them and destroy them. But this only confirms what was above, that bacteria are not objects of great importance as contagia or diseases affecting plants."

The validity of Smith's work on bacterial diseases was challenged in 1897 by Alfred Fischer, professor of botany in the Koniglichen Gymnasium at Leipzig, who had studied with De Bary in 1881. In a book entitled "Lectures on Bacteria" he denied the reliability of the results of Smith and others who claimed to have seen bacteria in plant cells. Smith replied to Fischer in 1899, pointing out his complete lack of sound knowledge of the subject. Fischer replied the same year, and Smith replied in two rebuttal papers, which seemed to end the controversy. While this debate is now chiefly of historical interest, it is significant that more than 20 years after the work of Koch and of Burrill a rather prominent botanist still challenged the existence of bacterial plant pathogens.

VIRUS DISEASES

About 1870, there occurred in the tobacco-growing region of Holland a disease of this plant which reached such damaging proportions that farmers sought the help of government scientists. Adolf Mayer, director of the agricultural experiment station at Wageningen, undertook a study of the disease about 1880. He published his report in Dutch in 1885 and in German in 1886. He was German-born and German-trained and was apparently acqtainted with the work of Koch. In searching for an explanation of the cause of the disease, he studied soil fertility, air-temperature changes, cultural methods, seed transmission, and nematodes but did not relate anyone of those factors to the etiology.

He described the disease, and, from the mosaic pattern common on leaves of affected plants, he suggested the name "mosaic," which has been retained until the present as the common name of the disease. Mayer discovered that when he macerated the tissue of a diseased leaf and injected the juice into the midrib of a healthy plant, the latter, after about 10 days, began to show typical symptoms. These appeared, not on the inoculated leaf, but on the youngest leaves of the growing plant, and as new leaves expanded, they were also diseased.

This is the first known record of the experimental transmission of a virus disease of plant or animal. However, Mayer, impressed by the ease of transmission, endeavored to find out more about the infective agency. The possibility of an enzyme was considered, but this seemed to him to be an untenable theory since he removed the infectious agent when he filtered the juice through a double layer of filter paper. Enzyme precipitates with weak alcohol were noninfectious. Continuous heat at 60°C. aid not alter infectivity of the juice, but several hours at 80° destroyed it. He applied the techniques of Koch" and others in an attempt to isolate a bacterium which could be shown to be the causal !agent. All bacteria isolated were noninfectious, and cultures of a number-I of known bacteria, when used as inoculum, gave negative results. Mayer 1 closed the investigation, still believing that a bacterium which had escaped I him would probably be found to be the infective agent.

About the same time that Mayer was conducting his studies of tobacco mosaic, Erwin F. Smith undertook an investigation of the peach-yellows disease, which had long been destructive in the United States. He was unable to transmit the inciting entity by transfer of plant juice as Mayer had done. He succeeded in transmitting it only by transplanting a bud from a diseased tree to a healthy tree, whereupon the latter eventually became diseased. He received Mayer's paper before he published his final results. He was of the opinion that the two diseases had much in common in the nature of the causal agent.

Dmitria I vanowski was a Russian investigator at the Botanical Laboratory of the Academy of Science at St. Petersburg. The mosaic disease had become troublesome is the Crime, and he upertook look with it about 1890. He confirmed Mayer's result with transmission of the causal entity by means of expressed plant juice. He destroyed the infectivity of the juice with heat.

He interpreted the causal entity as bacterium because of the lack of "fungi and other parasites." He did I not find, as Mayer reported, that filtration of the plant juice through a double layer of filter paper removed the causal factor, and he pointed out i that Mayer was in error in assuming that this procedure would exclude I bacteria from the filtrate anowski passed the infectious juice through a Chamberland filter candle and found the filtrate to retain infectivity. This fact is of great is torical interest because it is the first record of the I passage of either a plant or an animat virus through a bacteria-proof filter. This technique was so new that even I vanowski was skeptical, and he was apparently unaware of the great significance of the result.

He thought there was a possibility that the bacteria of the tobacco plant in some way went through the filter and made the filtrate infectious, although when he passed liquids which were favourable media for bacterial growth through the candle, they remained sterile for several months. Another explanation of the results, which seemed plausible to him in light of prevailing opinions, was that, a toxin secreted by the infectious bacteria in the sap passed through the filter and made the filtrate infectious. The work of I vanowski thus added one hew fact to those proved by Mayer, namely, that the infectious entity in juices from infected tobacco plants passed through a Chamberland filter.

The next investigator to approach this problem was Martinus Willem Beijerinck, Dutch bacteriologist, who was a colleague of Mayer at Wageningen in the early eighties ana moved to Delft in 1887. He took up a study of tobacco mosaic about 1897. He found the juice of infected plants, when filtered through porcelain to be sterile but still infectious. He sought anaerobic bacteria in the filtrate without success. A minute amount of filtrate was sufficient to inoculate several plants, and he demonstrated that the contagious entity was increased in some way in the infected plant. He called it a "contagium vivum £luidum" and referred to it subsequently in the paper as a virus.

Thus when Beijerinck published his results in 1898, he confirmed the findings of Mayer and of Ivanowski and claimed more emphatically than either of them that the causal entity was not a bacterium or any conceivable corpuscular material. He found that the virus would remain infectious in infected leaves which were dried down and kept in the herbarium for two years.

He regarded it, therefore, as capable of retaining its existence outside of the plant. He compared the tobacco virus with that of peach yellows and regarded the latter as a contagium capable of existing only in living plant tissue. In the same category as peach yellows he placed the plant variegations of Aubutilon and other ornamental plants which were transmitted to nonvariegated plants by grafting.

Erwin Baur published a paper on infectious variegation in Abutilon in 1904. Baur was a German botanist who later became prominent as a geneticist. He added little to the facts already known but confirmed the fact that. Abutilon variegation was due to a virus more like that of peach yellows than that of tobacco mosaic.

He maintained that the virus could not be an organism, since if a parasite had to depend upon man to make artificial grafts to perpetuate it, it would have long ago disappeared in nature. His argument is not sound in the light of present-day knowledge of insect transmission of.most viruses. At that time no one seems to have considered the possibility of insect transmission of plant Viruses.

The question as to whether the plant viruses were living or nonliving entities was debated for another 30 years. The events of those three decades will be discussed in the chapter on Virus Diseases. Suffice it to say here that at the beginning of the twentieth century the field of virus diseases of plants was just opening.

Two major types that we recognize today had been studied, i.e., tobacco mosaic and peach yellows. It was shown that the causal entities were transmissible and that they increased in the host. One was filterable through bacteria-proof filters. One was capable of existing for a time outside the living host; the other was not. Most of the present store of knowledge about plant viruses was still to be learned.

GROWTH OF PLANT PATHOLOGY IN THE UNITED STATES

Agriculture in the United States, !or 200 years after the establishment of the Colonies, was concentrated along the Atlantic seaboard east of the Appalachian Mountains. In the period from about 1830 to 1850 the opening of public lands in the Middle West was occasion for rapid expansion of population and development of agriculture in the;newly settled areas. Potato late blight appeared in the United States in the late thirties and was rampant throughout the forties at the same time it was destructive in Europe. As wheat growing expanded in the Middle West, rusts and smuts followed the crop.

Schweinitz collected and named many parasitic fungi in eastern United States from about 1820 to 1835. He was followed by Ellis and by Curtis. In 1875 Farlow returned from Europe with the stimulation of association with De Bary and at Harvard took early leadership in the study of parasitic fungi.

During the Lincoln administration, Congress passed the Morrill Act which provided land grants to each state, the proceeds of which I were to establish colleges of agriculture and mechanic arts. In the Federal I administration, matters pertaining to agriculture were first handled by the Commissioner of Patents. A Department of Agriculture was established in 1862, and it was first represented in the Cabinet by a Secretary in 1888.

Within the Department of Agriculture a Division of Botany was organized in 1885 under F. Lamson-Scribner. Within this division a Section of Mycology was formed in 1886, the name being changed to the Section of Vegetable Pathology in 1887. B. T. Galloway was head of the Section from 1888 to 1890: His early associates were Erwin F. Smith, Miss E. A. Southworth, and David G. Fairchild. Scribner, Galloway, and Fairchild carried out some of the earliest trials with the newly discovered Bordeaux mixture in the United States, especially on grapes, for the control of black rot.

As, the section grew, it was elevated to a Division of Vegetable Pathology in 1891, and to its staff were added A. F. Woods, N. B. Pierce, M.. B. Waite, W. T. Swingle, J. F. James, P. H. Dorsett, H. J. Webber, Theodor Holm, and M. A. Carleton. As tie work of the division was broadened, its name was changed in 1895 to the Division of Vegetable Physiology and Pathology. In 1901 the Bureau of Plant Industry was organized.

In the period from 1885 to 1900 plant pathologists in the U.S. Department of Agriculture carries out numerous pioneer studies. Among these were investigations of fungicides and studies on grape diseases both in eastern United States by Scribner, Galloway, and Fairchild and in California by N. B. Pierce. Smith studied peach yellows and rosette and later inaugurated his life work on bacterial diseases of plant. Webber and Swingle began studies of citrus diseases Florida. Smith opened up the work on vascular *Fusarium* diseases of cotton, cowpea, and water melon. Woods worked on nonparasitic diseases and on tobacco mosaic.

In 1887 Congress passed the Hatch Act, which made an annual appropriation of 15,000 to each state to found and support an agricultural experimentations. This support of agricultural research resulted in the appointment of a station botanist in many of the new experiment stations. Many of these men initiated programmes of research on plant, diseases in their respective states. Beginning about 1890, plant-disease research began in numerous centres in addition to that started in 1885 in the U.S. Department of Agriculturc.

PLANT PATHOLOGY IN 1900

A hundred years after Prevost, at the opening of the twentieth Century, we find a new science of plant pathology well established in Europe the United States. During the first half of the nineteenth century both animal and plant pathology struggled to define and gain recognition for the germ theory of disease. The lack of adequate techniques to study the life cycles of fungi and bacteria greatly impeded the advance of knowledge of these lower organisms as pathogens. To Brefeld and to Robert Koch goes the major credit of developing methods of pure culture of fungi and of bacteria, respectively. The last quarter of the century saw rapid adoption of these new tools and a corresponding increase in volume of new information on plant diseases.

The last half of the nineteenth century was predominantly a descriptive were in which many new species of pathogenic fungi and a few of pathogenic bacteria were described. Emphasis was laid upon the naming and describing of causal organis and correspondingly less.attention was given to other phases of disease. Consideration of the influence of environment on disease, which was usually regarded as the sole cause of disease before the De Bary-Kuhn period, waned decidedly in this eras Control measures received impetus with the discovery of. Bordeaux mixture, and during the last decade of the century growing attention was given to new fungicides for spraying of foliage and for treatment of seeds.

By 1900, bacteria as plant pathogens had been finally accepted. Virus diseases had been recognized as a distinct type of infectious malady, but most knowledge concerning them was still to be attained.

In the twentieth century, phytopathology has expanded so far and so rapidly that we can undertake to mention here only the major lines of development. This period will, in the future, be distinguished by a broadening of plant pathology into a well-rounded science which in its interrelation with other fields embraces the study of plant disease in its broadest aspects. There may be mentioned several of the major lines of emphasis.

The relation of environment to disease development its influence upon the host, upon the parasite, and upon the interaction of host and parasite. The nature and inheritance of variability in the pathogen with particular reference to its pathogenic property. Virus diseases; the chemical nature and the biological activity of plant infecting viruses. The nature and inheritance of host resistance to pathogenic organisms; development of disease-resistant varieties of crop plants. The improvement of fungicides and of methods for estimating their fungicidal and fungistatic properties. The prevention of disease through regulation, quarantine, inspection, certification, and education.

3

Fungi as Plant Pathogens

INTRODUCTION

Plant diseases are due mainly to the action of fungi, but some are caused by bacteria, a few by plasmodia, while others in which no particulate living organisms have, so far, been discovered, are believed to be due to virus infection. In plant viruses that have been studied in any detail the invisible agent multiplies with great readiness only in the living host from which the virus has been isolated in the form of a nucleoprotein. Other diseases, again, but of a non-parasitic nature, are attributed to physiological or to deficiency troubles, and probably all are influenced by the factors of the environment.

Including for the moment the bacteria with the fungi proper, one finds an enormous assemblage of species, at least as many as in the flowering plants, having few characters in common. They may be unicellular or composed of many cells, thread-like or forming sheets or masses, shapeless or beautiful in design, of limited or practically unlimited growth, of many colours. In the lowest group a cell wall- that standard criterion of the plant cell- may be present only in the sporing stage. Motility is not absent, either as the slow movement of plasmodia or the swimming of zoospores or the flagellar movements of many bacteria. In structure some approach closely to the green algae and, as the name Phycomycetes or algal-fungi reveals, it was long a popular suggestion and is still sometimes upheld that they were derived from algae; the best modern view, however, is that they arose from colourless unicellular protozoa, and it is even maintained that, since the green algae probably arose from similar protozoa that had acquired chlorophyll, and since the rest of the green plants have developed along that path, it is logical to regard the fungi as a distinct phylum, neither animals nor plants but co-equal in their origin with the green organisms forming the plant kingdom. Normally, however, they are treated as plants, included in botanical works and courses, and they may be accepted as such.

The chief characteristic of the fungi is the absence of chlorophyll, even in those members which happen to be coloured green. This carries with it a profound difference in mode of life from other plants, for they take their food in the form of organic matter, whereas the chlorophyllous plants obtain their carbonaceous food from the carbon dioxide of the air, through the energy of sunlight. The flowering plants, ferns, mosses, and algae build up their food out of dissolved inorganic salts from the soil or water, together with the constituents of the air.

The fungi are unable to do this; they are incapable of photosynthesis and so cannot make carbohydrates from carbon dioxide and water; they must have organic matter or food already prepared by having been built up into the bodies of plants and animals, or having once formed part of these, like jam or leather. They are scavengers, without whose cleansing aid in decomposing organic matter, the earth would soon become unfit to sustain the higher forms of life.

BACTERIA

The bacteria really have very little in common with the fungi proper, except, in general, their mode of feeding. They are extremely minute, usually consisting of a single cell or a collection of more or less independent units. Sometimes these units are arranged in branching threads which are hard to distinguish from those of the filamentous fungi. The fungal thread, however, is normally a part of the one individual, whereas a bacterial thread consists of a chain of individuals more or less loosely joined together. The branching is also different, the fungi having true branches, while in a bacterial column the spurious appearance of branching is due to imperfect separation of broken parts of the chain, which continues to give off new individuals above and below the break.

The most important distinctions between the two groups are in the methods of growth and reproduction. Fungi normally grow only at the ends of the threads which compose their bodies, and multiply by forming spores at the ends of, or within special threads. Bacteria, on the other hand, have not got this purely end growth, and usually multiply by each cell simply' dividing into two cells, which frequently separate from each other and become independent.

There are other less important differences as in the nuclei, cell wall, and so cl Being smaller and lighter than most spores of filamentous fungi the bacteria are even more efficient than the former in reaching and decomposing organic waste; as f§ well known, they were the chief subject of Pasteur's convincing demonstration against the old-fanged notions of spontaneous generation.

PLASMODIOPHORALES

The Plasmodiophorales are an aberrant group having apparent affinities with the chlorophyll-free mycetozoa or myxogastres, organisms close to the dividing line between animals and plants. They are now usually placed near or in the primitive Chytridiaceae. They have a power of spontaneous creeping movement not limited to the zoospores. Instead of a germ-tube the spore gives rise as a rule to motile swarm cells which fuse to form a small mass of naked protoplasm endowed with movement resembling that of an amoeba.

This forms a multinucleate' plasmodium' or plasmatic body still without a wall and usually constituted by the amalgamation of several individuals. In at least some of the genera the swarm cells function as gametes and the plasmodium arising from the zygote forms sporangia, so that there are well-marked sexual and asexual phases in the life-cycle. Two important crop parasites binning to this small family occur in the British Isles, namely Plasmodiophora brassica and Spongospora subterranea, causing' club root' of cruciferous crops, and' powdery scab' of potatoes, respectively.

THE FUNGI PROPER

All known fungi, with few exceptions, originate from spores, bodies comparable with, though not strictly similar to, seeds, for they have no preformed embryo or germ. These spores germimtte, requiring for the purpose much the same conditions as ordinary seeds.

The result of germination is usually the protrusion of one or more fine filaments known as germ-tubes. These grow and give off branches, the resulting branched thallus constituting the' mycelium' or vegetative part of the fungus, as distinct from the reproductive part. Each individual filament of the mycelium is termed a 'hypha'. In a limited number of primitive fungi the vegetative part consists of a single cell, usually round' or oval, and this may become transformed into a reproductive cell with little change in shape; the life-history being a very simple one.

THE VEGETATIVE STRUCTURE OF FUNGI

THE MYCELIUM

At an early period in its growth the germtube and the hypha which arises from it usually become segmented into a row of cells by the formation of transversely walls each, when young, being perforated by a central pore sufficiently wide to allow protoplasmic granules and even nuclei to pass from cell to cell. Each cell is a hollow structure, bounded by walls of a transparent pliable material, which is sometimes cellulose as in the higher plants, but more often is a complex nitrogenous substance which resembles chitin, the material which forms the covering of insects.

Within the walls, the cell is filled with a watery jelly, protoplasm, a material mainly protein in constitution, found in all living bodies and, in fact, the actual living matter itself.

The cell cavity is not uniformly filled. Here and there little spaces occur and are known as 'vacuoles'; they are usually filled with water containing dissolved salts, sugar, and so forth, of the cell sap. Here and there also condensations of a peculiar kind of protoplasm form, which are termed nuclei. One or more of these nuclei may occur in each cell; two are habitually found in certain phases of the life-cycle of Ascomycetes and Basidiomycetes, playing a part in the complicated sexual reproduction of these fungi which will be further discussed below.

In the lower fungi it is common, or even in some families habitual, to have no transverse walls in the active hyphae, which are' coenocytic "that is to say, composed of a single large cell with many nuclei. The nuclei are the bearers of the hereditary characters of the cell and also play an important part in regulating its metabolism.

GENERAL PRINCIPLES

The plasm with its nuclei but many other substances, such as reserves of sugars, fats, glycogen, crystals, resin, pigments, and so on. They are of many shapes, but on the whole the spores preserve considerable uniformity within the species. The prodigality of nature in its reproductive schemes is illustrated by the fact that a single giant puffball has been estimated to bear some 100 million spores. It is little wonder that every substance suitable for colonisation by fungi is reached by air currents charged with spores, unless special precautions are taken to exclude them.

The body of the simpler fungi usually consists of branching hyphae only. It may be short or long lived. The' fairy ring' fungi, of which Marasmius oreades is the commonest in England, are the best examples of perennial mycelium, as they may survive for as many as four hundred years, producing a crop of toadstools each year. In many cases unions termed' anastomoses' occur between neighbouring hyphae which may be from the same or different spores of the fungus concerned. These form bridges across which nutrient material and even sometimes nuclei pass from one hypha to another. Most fungi, however, have that part of the body which is involved in spore-bearing of more complex structure, formed of hyphae of special shape, or of bundles of hyphae, or of an interwoven tissue derived from hyphae by their repeated division into short cells. This tissue is termed' plectenchyma 'or, since it resembles to some extent the parenchymatous tissues of higher plants,, pseudoparenchyma'.

CHLAMYDOSPORES AND GEMMAE

Hyphae of the ordinary vegetative mycelium may also become modified. Some of their cells may become provided with thickened walls, able to withstand adverse conditions, and may even assume a characteristic shape and colour. They are known as chlamydospores and, like true spores, are usually microscopic.

As they become separated from the mycelium when mature, and can germinate like spores, they playa part in the dissemination of the species which, on account of their durable nature, may be very important. Scarcely to be separated from the chlamydospores, except in being less durable, are the cells termed gemmae. They are usually thin-walled and contiue to bud out cells similar to themselves instead of giving new hyphae.

Under certain conditions the mycelium may be almost wholly transformed into these masses of rounded cells. Indeed, the familiar group of the yeast fungi may be regarded as an extreme case of the reduction of the mycelium to small gemmae endowed with a remarkable facility for budding out similar cells so as to form colonies which readily break up into their component units. This budding mycelium is frequently associated with fermentative activity.

SCLEROTIA

Much larger and more complex masses of vegetative hyphae-than those producing chlamydospores may become united into, sclerotia', sometimes measuring several inches across and as hard and heavy as a stone, but still formed wholly from vegetative hyphae variously thickened and divided so as to lose their filamentous character, and become pseudoparenchymatous. The nests of white ants in hot countries often contain numerous rounded to spindle shaped stalked sclerotia belonging to the Ascomycete Xylaria injuries while a sclerotium which may be as big as a toy football is formed by the basidiomycete Polyporus mylittae in Australia, where it is known as 'black fellows' bread', as it is eaten by the aborigines.

Several other idiomycetes also form large sclerotia. Similar but much smaller types are known in both hot and temperate climates, the commonest in the British Isles being the often irregularly shaped sclerotia of Botrytis or Sclerotinia, of black-dot of potato and tomato Colletotrichum atramantarium and of Corticium solani, the cause of, black scurf' of the potato, etc. The round sclerotia of Sclerotium rifles, resembling very small seeds, are derived from a parasite destructive in the United States and elsewhere, while the still smaller, almost microscopic, Sclerotium stage of the pycnidial fungus Rcrophomina phase oli is ubiquitous in India and known in several other hot countries.

In some sclerotia, such as those of Botrytis or Sclerotinia, a hard outer rind or cortex of thick-walled, often dark cells, with small cavities, may be distinguished from a softer pith. Besides being useful stores for reserve nutrient material, sclerotia are particularly well suited to withstand unfavourable extremes of temperature, dryness, and the like, and are valuable in carrying the organism over long periods of adverse conditions.

RHIZOMORPHS

Visible strands or cords of mycelium can often be seen on turning over heaps of leaves or at the base of puffballs and other large fungi. The simplest of these are generally white, and beyond being intertwined or sometimes laterally united they do not appreciably differ from an ordinary bundle of vegetative hyphae. More complex strands, known as, rhizomorphs "are found in the vegetative thallus of some fungi. In the less differentiated types of these, ordinary filaments unite with specially tough hyphae and sometimes with hyphae adapted for the conduction of nutrients to constitute a part of the plant fitted for extensive exploration in search of new colonisation grounds. In the dry rot fungus Merulius lacrymans they have been found passing through the mortar of walls to reach fresh structural woodwork on the other side.

The more highly differentiated rhizomorphs are represented in the common parasite of tree roots Armillaria mellea. These have been likened to purplish-brown or black leather shoestrings, by they are often much branched, rather uneven in diametre, smooth and shiny when fresh, and luminous when young.

They consist of an outer black, rather brittle rind, composed of brown, thick-walled pseudoparenchyma, and a light-coloured tough flexible pith in which the filamentous character of the hyphae is largely preserved. Though they grow normally by definite apical growth, keeping their root-like form, they can spread out into sheets of mycelium in certain positions, especially when they pass under the bark, or they may burrow deeply into the tissues of the root, forming tubular structures in the long axis of the root; these are still differentiated into a harder cortex and a loosely interwoven pith, and grow only at the apex.

Similarly embedded rhizomorphs are formed by the Ascomycete, a widely distributed: Ingus causing root rot of tea and other crops. In the more-serious of the root diseases of tropical plantation crops, such as rubber and tea, the rhizomorphs, which are the chief organs of spread, remain on or in the roots or other buried timber and even in Armillaria mellea their ability to grow away from such supports and food-base is limited, and extension through firm soil, other than along wood, is almost negligible. Their function in the parasitism of the last-mentioned fungus is discussed in a later chapter.

Allied to rhizomorphs are the raised black lines or narrow bands running longitudinally on the surface of roots attacked by Armillaria mellea, each representing the free margin of a dingy white sheet of mycelium which disrupts the wood, often but not invariably in a radial direction. The xylostromata are capable of growth at their free edges and of causing infection when a diseased root comes into contact with a healthy one. They are much less often seen in temperate than in tropical forms of the fungus, in contradistinction to the rhizomorphs, which are common in temperate, but may be hard to find in tropical countries.

The xylostromata arise by proliferation from a narrow band of deep brown cells which form sharply defined black lines in the tissues invaded by Armillaria mellea, These bands are sections of continuous sheets crossing all the tissues of the host from near the surface to the pith. They begin by a development of thin-walled, swollen, colourless, bladder-like cells completely filling the host cell cavity. The bladder cells soon thicken their walls and turn brown.

When fully developed, the black line is composed of a dense mass of angular, often rather small cells of a deep brown colour, and this colour also stains the walls of the host cells. The result is a continuous layer surrounding an area rotted by the fungus, and it has been suggested that it encloses a mass of fungal tissue, mixed with elements of the host, which is protected from adverse influences and to which the term '*pseudosclerotium*' has been applied. Black lines of this type occur in the tissues invaded by various parasitic and saprophytic species of Ustulina, Xylaria, Hypoxylon, and the like.

STROMATA

In many fungi, after a period of vegetative growth in the filamentous condition, the mycelium condenses here and there into pseudoparenchymatous masses termed' stromata "which are formed generally on or just below the surface of the substratum in which the fungus is growing.

These differ from lerotia only in having usually a less regular shape and a margin in more frequent communication with the filaments of the rest of the thallus. They often, indeed, as in Rhytisma, only form flat crusts covering the deeper hyphae from which they have developed. They commonly constitute a transitional stage between the purely vegetative and the reproductive parts of the fungus, and the term, storm is sometimes restricted to this pre-sporing stage. It is difficult, however, to accept this restriction, as stromatic crusts are common that seem to have no association with spore bearing.

The parts of the fungus plant so far considered are chiefly concerned with nutrition. The more specialised rhizomorphs, sclerotia, resting and budding cells, take part also in propagating the species, but this is 'vegetative' propagation, on a par with propagation by cuttings, buds, suckers, and the like, amongst higher plants. The reproductive system proper is usually quite distinct and well defined.

REPRODUCTION OF FUNGI

SPOROPHORES

As the vegetative system is usually buried in the substance on which the fungus feeds- in the soil, in rotting organic matter, or in living plants- and as most spores should be formed in free access to the air to secure their dissemination, it is evident that some special arrangement must be made to bring the spore-forming parts where they will be exposed to the elements.

This is done usually by the development of structures known as sporophores raised vertically above the level of the general mycelium. The sporophore may consist either of a single hypha or of bundles or complex masses of hyphae and it bears on some part, usually towards the top, the swore of tnfungus. The hyphae of which it is corpuscle ordinarily differ from those of the vegetative mycelium in one or more characters, such as their vertical position, limited growth, or special structure and as "they may be arrange a in systems of great complexity, an almost endless series of different types of sporophore exists.

The simplest Iotina-commonly 'in riiany of the lower fungi and in those moulds that are familiar on damp or rotting organic substances and that belong to several different families of the fungi. In some of these, particularly such as have a considerable development of aerial mycelium in addition to the hyphae buried in the substratum and some of those that grow in water or in soil, special sporophores are almost or quite dispensed with, as being unnecessary to secure free dissemination.

The spores are formed at the tip of, or laterally on, ordinary hyphae, or, in the aquatic species, in special cells on the hyphae, and are liberated in the air, the water, or the soil as the case may be. More often, however, specialised sporophores are formed, and in the simplest cases these are single hyphae called sporangiophores, or conidiophores, often of distinct shape and size, which grow out into the air, and bear the spores on the tips or sides or on lateral branches.

In other cases, clumps of hyphae growing close together rise into the air. Sometimes these merely remain near together without lateral union, sometimes they unite more or less closely into solid columns.

Sporophores aggregated into crusts or layers are produced in many fungi. Sometimes these are formed on the surface of the substratum in which the vegetative mycelium grows, sometimes, especially in parasitic and other fungi that grow in the stems and leaves of plants, just below the surface, which is eventually ruptured so as to expose the spore-bed. In either case, the layer of sporophores may arise directly from the mycelium below, or there may first be formed a stromatic tissue composed wholly or in part of pseudoparenchyma. Common examples of spore-beds formed below the surface of the plant are the, sari' of the cereal and other rusts and the 'acervuli' of many of the Ascomycetes and of the 'Fungi Imperfecti'.

When formed on the surface, there may be a very large mass of stromatic tissue intervening between the vegetative mycelium and the layer of sporophores. Such compound sporophores in the form of superficial stromatic crusts or tubercles may be found commonly on pieces of old wood, dead branches of trees, and the like, in damp situations. Sometimes the edge of the crust becomes free and curves over to form a sort of bracket. This is the first stage in a process which has as its purpose the lifting of the spore-bearing part of the sporophore above the object on which it is growing. In more advanced types, so much of the crust may leave the surface that only a narrow band or stalk remains to join it to the latter. In this way the bracket fungi found on trees and old wood, and the laterally attached toadstools, are formed. Finally, the whole spore-bearing part of the fructification may be raised from the surface on a central stalk and may be expanded into a cap at the top, as in the well-known mushrooms, or may remain as a simple or branched, vertically elongated body of various shapes.

In another type of compound sporophore, the spore-bearing hyphae are enclosed in roundish or cup-shaped or flask-shaped receptacles. To a common form of this type the name pycnidium 'is applied when it bears spores on the or immersed in it) e.g. Epichloe, Claviceps and opening to the surface by the narrow neck through which the spores are expelled when-ripe.. Sometimes they are formed close together but free from each other as in Nectar and Ophiobolus, sometimes in groups immersed in a 'common stroma as in Claviceps and Epichloe, and, as in these two types, the stromabas in some cases become eleyated or vertically elongated so as to raise the spore receptacles above the level of the sub stratum.

In most compound sporophores, the spores are not produced on all parts of the organs. In the mushrooms, for instance, they are formed only on the surface of the gills found on the under side of the expanded 'cap'. To the fertile part of the sporophore, i.e. that part which actually bears spores, the name 'hymenium' is given.

In many cases, where the spore-being hyphae are so numerous as to form definite layers or a hyrpenium, certain sterile hyphae usually of characteristic form, occur intermingled with the fertile ones. These are frequently known as, paraphyses' and are especially common in the Ascomycetes and Basidiomycetes. The basidia of the H ymenomycetes are often interspersed with sterile cells or 'cystidia'.

From the above account of the fungus body, it will be seen that an ordinary fungus consists of a vegetative part, usually formed of thread-like hyphae buried in the substratum, find of a reproductive part which is sometimes not distinct from the vegetative except in bearing spores, but is more usually modified to form a simple or compound sporophore. As the production of spores provides for a new generation of the fungus, one may consider that the life-history of any species begins with the germination of the spore and ends with the development of new ripe spores, or in the case of those fungi which have more than one kind of spore with the development of the final or, as-it is often termed, 'perfect spore form, which is usually the result of; sexual process at its equivalent.

FORMATION OF SPORES

The spores of fungi present a wonderful range of variation, there being hundreds of types differing in size, snaps, structure, and more of formation. The one fungus may have several different kinds of spore during the course of its complete life history. The vast majority differ sharply- from the vegetative cells of the mycelium, but in any single species each kind of spore remains remarkably constant. Two main types may be distinguished: a 'perfect' form which is the result, immediate or delayed, of a sexual process and an 'imperfect' or 'asexual' from to which the term sporangium or conidium' is usually applied. When the thallus which bears self capable of sexual fusion is distinct from that carrying- asexual spores, the former is termed the gametophyte and the latter, the spordphyte. The best example of this alternation of generations amongst the fungi' is found in the genus Allomyces but it is much less clear cut in fungi than in such plants as mosses and ferns.

SEXUAL REPRODUCTION

The perfect form of fructification differs widely in the three main groups of fungi, Phycomyceies, Ascomycetes and Basidiomycetes. In the Phycomycetes it is a thick-waned, well-defined cell, to which the general term 'zygote' is applied in some of the more primitive forms, while in the Oomycetes the zygote is termed apoospore and in the Zygomycetes, a 'zygospore'. A zygote is simply the end product of the fusion of two cells with sexually differentiated nuclei which also fuse. These cells are known as 'gametes' and in some of the Chytridiaceae are motile; they may be outwardly alike, so that a IV-ale is not to be distinguished from a 'female' gamete.

An oospore is a cell resulting from the fertilisation of a larger female' cell or 'oogonium' by a smaller male cell, the' antheridium "both borne on hyphae of the mycelium. A zygospore results from the union of the tips of two branches of the mycelium, often quite similar to one another. In all these cases fertilization is completed, as usual, by the fusion of the nuclei from the two uniting gametes to form the zygote. At some stage prior to fusion the nuclei undergo a 'reduction 'division, which results in the halving of the number of 'chromosomes' and gives rise to what is termed the I, haploid' phase of the life-cycle.

When two haploid nuclei come together in a cell, the dikaryophytic, dikaryo-, or dikaryotic phase is initiated; fusion of the two haploid nuclei restores the original number of chromosomes, half of which are now of different origin from the other # half, and the I resulting nucleus is termed 'diploid'. The process directed to bringing together I the two haploid nuclei in one cell is sometimes called 'diploidisation'.

In the Ascomycetes the perfect stage is the 'ascus', a saclike, roundish or elongated cell containing' ascospores "which is sometimes immersed without orderly arrangement in a stroma or in meshes of the mycelium but usually lies side by side with other asci and, perhaps, paraphyses to form a definite humanities. In some forms the ascus is the result of the fertilizations of a female cell, the' ascogonium.', through a process or 'trichogyne' which contacts an antheridium-like male cell.

Instead, however, of the two nuclei uniting as in the oogonium,,they- may divide several times and a male and a female pair pass into each of a group of cells which grow out from the ascogonium; from one or more of these cells the asci 'arise, each ascus still having a male and a female nucleus. Fusion of.the two nuclei occurs in the young ascus and the ascospores arise by the division of this sexual nucleus. Owing to their relative sizes the nuclei often have to lie in single file in the ascogenous hyphae: and an arrangement known as the 'crosier' of nuclei, male and female, in the penultimate cell, from which the ascus arises. The clamp Connection of the Hymenomycetes, to be described below, serves the same purpose.

There is still some disagreement as to the number, location and significance of the nuclear fusions preceding the formation of ascospores and the details of the sexual process are as yet uncertain in some forms. In some cases a fusion of the male and female nuclei has been observed in the ascogonium, in addition to the fusion in the young ascus, and two reduction divisions following this double fusion have been described.

In the unicellular yeasts the single cell may become converted as a whole into an ascus, the formation of which may be preceded by union with another cell by means of a short bridge across which the nucleus,of one cell passes to face with that of the other. In cultures of-these fungi some cells are haploid, some diploid. The great bulk of the Ascomycetes are contained in' two main groups, the Pyrenomycetes in which the sporophore is a flask-shaped perithecium often immersed in the tissues or in a stroma and opening by a narrow neck and the Discomycetes in which the hymenium lines the inner surface of a cup-shaped sporophore or apothecium. In the Pyrenomycetes, which are the most numerous and diverse group of Ascomycetes, thc hymenium lines the inner surface of the perithecium and the spores may se extruded while they are still enclosed in their asci or after they have been set free into the' cavity of the sporophore.

In the heterogeneous group, the Plectomycetes, the powdery mildews, are furnished with a characteristic type of fructification; the cleistocarp or cleistothecium, which is not provided with opening and the ascosporesare only liberated after the protective wall ruptures or decays. The Exoascaceae are included in this group although the asci are not enclosed in a special body at all, but like Taphrina deformans the cause of peach-leaf curl are exposed on the surface of the host once they break through the cuticle at maturity. The typical number of spares in an ascus is eight but there may be fewer or more, when the number is increased by division of the original nuclei, usually to a higher power of 2, as 16, 32,-.64, 128, but sometimes irregularly from suppression, of one or more spores.

In many of the higher Ascomycetes the ascogonium which eventually gives rise to the ascogenous hyphae appears often to be fertilized by microconidia. The oidia may be borne laterally on small awl-shaped or verticillate sporophores and in certain fungi as in the genus Sclerotinia do not function as ordinary spores as they do not give rise to a mycelium. When set free in Sclerotinia they adhere to cells either in special receptive bodies raised above the level of the stroma or contained in the sclerotia which are characteristic of these fungi. Fusion presumably results in the transference of a nucleus from a hypha in one mycelium to another, and apothecial development follows.

Spermatia borne in special spermagonia have long been known to occur in the life-cycle of many rusts. Pyrenomycetes and Discomycetes and to have no apparent function as sprees capable of reproducing the species. In a few cases they have also been found fusing with cells of the primordia of pedthecia, and though the evidence is largely circumstantial, it is probable that both oidia and spermatia ordinarily function as haploid' male' cells.

Turning to the Basidiomycetes, a still greater complexity is found in the processes which seem to.serve the ends of a sexual union. In place of the ascus of the Ascomycetes, the 'basidium' marks the perfect stage in the life-cycle. The term' promycelium' is also applied to the basidium in this group. It is a round, club shaped or baton-shaped body in which the gametic nuclei fuse. Subsequently it gives out a group of spores on short outgrowths from its surface. In the wide sense, the Basidiomycetes include the smuts and rusts as well as the mushrooms and bracket fungi, and the different groups vary considerably in the details of their perfect stages since the' sporidia of the two former groups do not arise immediately from the zygote.

The sporidia of the smuts are borne typically on a promycelium which resembles a stout germ-tube arising from the smut spore. In the majority of the species the promycelium is divided by transverse septa into four cells each of which contains a single nucleus. Usually a single sporidium or hyphal branch is produced from each promycelial cell, but sometimes two or more are developed; the promycelial nuclei do not pass out into the sporidia but divide, and each of the daughter nuclei enters a sporidium.

In some cases this nuclear division may be repeated several times so that crops of sporidia arise directly from a promycelium. Sometimes the four cells of the promycelium use in pairs by means of small buckle-like outgrowths through which a haploid nucleus passes from one cell to the other. Thus the binucleate, dikaryotic condition in the smuts may arise from the union of two sporidia or hyphal branches or from the union of two promycelial cells. In the genus Tillejia, the nuclei usually pass into the sporidia or hyphal branches, none remaining in the promycelium which in this type usually remains uniseptate.

In Tilletia the primary sporidia formed at the tip of the promycelium become binulceate by fusion in pairs; they give rise to fine hyphae on which are borne, secondary sporidia which are usually also binucleate but on germination give uninucleate and presumably haploid sporidia. These secondary and tertiary sporidia are considered by some to be the true basidiospores. The smuts are parasites, and in a good many species it is the binucleate germ-tubes which enter the host plant, the cells of the mycelium in the host usually again appearing binulcleate just before the smut spores form, when the two nuclei fuse, transforming the smut spores into diploid zygotes.

In some species the fusion hyphae are formed within the tissues of the cost and there is some uncertainty as to the details of the change from the haploid, to 'the diploid phase in these. The completion of the sexual process from the time the two nuclei are brought together in a cell until their fusion in the young spore may take several months. The sporidial stage can be readily grown in culture and experiments with cultures derived from single sporidia have established that fusions will only occur between certain pairs and that the four promyceliaf cells, usually represent two males and two females so that a simple form of heterothallism exists., Since in some species more than one pair have been found capable of fusing it is necessary to recognise the existence of more than two kinds of sexually functionable gametes in fungi or else to use another terminology. Thus some speak of the 'A', 'B', 'C', etc., 'genders' instead of sexes, or of compatible and incompatible groups. At least nine such groups appear to be recognisable in maize smut.

It has Jong been known that the rusts have a dikaryophytic phase in which two nuclei occur in each cell. This commences as a rule somewhere in the primordia of the acedia and terminates in the ripe teleutospore where the two nuclei fuse to form the zygote. Rusts are amongst the most pleomorphic of the fungi, the full life-cycle often possessing no less than five spore stages. The basidium arises from the germination of the teleutospore—a resistant spore-stage. Homologous with the smut spore and haploid phase is initiated during the process of nuclear division in the promycelium, and which normally results in the production of four sporidia one from each of the four cells into which the promycelium becomes divided by trans verse septa.

As the four sporidia belong to two sexes. The haploid sporidia infect the host and cause the development of immersed flask-shaped spermagonia containing masses of tiny spermatia which are extruded in a drop of sweet 'honey dew' or nectar through the neck or pore to the surface of the plant. Here they are visited by flies or other insects which carry them to the openings of other spermagonia where they adhere to certain emergent hyphae in or at the margin of the spermagonium.

The next stage is difficult to follow and seems to differ in different species. In some, at least, the nucleus of the spermatlum appears to be able to pass back through the hypha with which it fused to reach the deeper levels of the haploid mycelium developed from the sporidium. The general outcome is the establishment of a binucleate condition of certain haploid cells at the base of the future sporophore which develops in the vicinity of the spermagonia, often on the opposite side of the leaf. This is the aecidium, bearing binucleate aecidiospores in chains.

From the dispersed aecitliospores arises a parasitic mycelium, the paired nuclei in the cells of which are maintained by conjugate division. It is this dikaryophytic mycelium that gives rise to binucleate uredospores, often in repeated crops of uredosori in such quantity that the yellowish or rusty spores colour the surface of the soil under the infected plants. Ultimately teleutosori replace the uredosori and the paired nuclei at last fuse in the mature teleutospores.

The rusts frequently overwinter in the teleutostage, producing their promycelia when the teleuto spores germinatein the spring. In a considerable number of species the haploid sporidia from the promycylium are alliable to infect the species of host plant on which the uredo-teleutospore stage develops, but attack some other species, often far removed systematically from the first and cause the development in it of a haploid vegetative mycelium leading to the aecidial stage while the binucleate aeeidiospores carry the fungus back to the original host. This which is known as heteroecism will be further discussed in a later section and seems to be a step in advance along the path already entered by the smuts, where the haploid germ-tube may be devoid of parasitic capacity.

In the Basidiomycetes proper the conception of sexuality must be further stretched if it is to cover all the known cases. Not only are the four sporiqia or basidiospores that are normally borne on four sterigmatabn the free end of the basidium often only able to form thalli that can fuse in pairs, which are mutually interested, but isolation of sporidia from other sporophores or other regions may go on adding almost indefinitely to the number of sexually functional gametes found. In these fungi the haploid spore stage may be represented by uninucleate oidia horne laterally on the mycelium, which may fuse together or with the end cells of hyphae composed of uninucleate cells. The result is a binucleate cell and the binucleate condition in subsequently developed cells of the terminally growing hypha is maintained by a curious contrivance called a 'clamp connection.

This, whose presence marks the dikaryophytic phase of the mycelium, develops around the September. I turn formed when the end cell of the hypha divides. The clamp begins as a protuberance in the side of the end cell just above I where the new septum will form Both nuclei move to this point and one enters the protuberance. Both then qivide while the septum separating the parent cell into two develops between the pair arising from the nucleus which did not migrate into the. protuberance.

The clamp completes its development by uniting at its tip with the cell below the new septum, and a nucleus from the pair formed in it passes to each of the two cells of the hypha. Thus the binucleate cells always contain a nucleus from each of their two components. Eventually the two fuse in the basidium and the perfect stage in achieved. Some times the dikaryophytic mycelium arises from the fusion of haploid mycelia derived from oidia of opposite sex as well as from haploid mycelia of.opposite sex derived from the germination of the basidiospores.

These haploid mycelia are without clamp connections and are sometimes termed primary mycelia as distinct from the' secondary, mycelia of the dikaryophytic phase. Oidia are some times developed on secondary mycelia, when they are usually binucleate and function as ordinary conidia. There appear to be many interesting variations in the diploidisation process, even to the transfer of nuclei from a diploid to a haploid mycelium, but it would be outside the scope of this survey to follow these or the debates to which they have given rise as to the physiological significance of the nuclear behaviour.

There have been several reports of clamp connections in the smuts where the paired nuclei cell is less rigidly preserved during parasitism in the rusts. It seems that both in smuts and Hymenomycetes the partnership of the two nuclei is liable to dissolution. In Urocystis cepulae the parasite of onion smut, the cells remain long uninucleate in onion tissues, becoming dikaryophytic just before sporulation.

HETEROTHALLISM

In all the main groups,of fungi it Lias been found that individual thalli or vegetative bodies are not necessarily identical in all their attributes in any given species. The difference to which the term' heterothallism 'is applied relates to sexual behaviour; in 'homothallic' forms thalli derived from a single spore can give the perfect stage, whereas in heterothallic species.

The perfect stage is only produced when union occurs between the cells or gametes of certain pairs of thalli. Thalli that can contribute to the development of the perfect stage are sometimes termed 'plus' and 'minus' thalli but it is scarcely possible to restrict the conception to a single pair of signs. It is evident that heterothallism is a means of securing outbreeding, a matter the biological value of which is indicated by the many well-know adaptations directed to secure the cross-fertilization. The apparent capacity of the different strains inheterothallic fungi to maintain themselves indefinitely in the imperfect stage, so that there is no way of knowing, that they have the ability to produce the perfect form unless chance or deliberate experiment reveals it, suggests that some at least of the great mass of the Deuteromycetes or 'imperfect fungi' may simply he strains that have lost their mates.

An alternative suggestios that nutrition rather than sexuality forms the basis of heterothallism has found some support. In this nuttitive theory each of the various strains is thought to possess a gene or factor absent from, the others; enabling it to utilise some essential food material. Union of compatible strains brings ability to use all that is required.

ASEXUAL REPRODUCTION

Out of the multiplicity of asexual spore forms found in

the fungi certain main types may be mentioned. The 'sporangia' of the Phycomycetes maw contain unicellular 'zoospores', the animal-like attribute implied in the name being their power of swimming about in water by the movement of one of two 'cilia' or 'flagella', oarlike protrusions of the naked protoplast, attached terminally or laterally. After this free-swimming period the zoospore comes to test, rounds off, usually becomes provided a cell wall, and then germinates by a germ-tube or by a short process which penetrates the host cell. Germination sometimes results in the liberation of a second zoospore on which the position of the flagella may now be different, apical or lateral, from their previous position, so that the period of free-swimming life is pro longed.

A transitional stage in the adaptation of the aquatic fungi to life on land is, perhaps, to be observed in the Peronosporaceae and some of their allies. In many of these the sporangium often fails to produce zoospores but germinates directly by one or more germ-tubes, just as if it were wholly transformed into a single spore. In the Zygomy cetes, unlike the aquatic families of the Phycomycetes, the spores borne in the sporangia are not motile but are suitable for passive dissemination in the air or water; these are amongst the commoner land moulds and include the well known Mucoraceae such as the bread moulds and others found on decaying organic matter in the soil and elsewhere. Most mucors have mucilaginous spore heads and are not readily dispersed until the head breaks up in water or in the soil solution and the individual spores adhere to soil particles and the like and are dispersed on.these.

In some of the other genera these dried minute unicellular spores, when liberated by the disintregation of the sporangial wall, become air borne and are blown about as dust. Pilobolus has the whole sporangial head shot by a water jet generated in the sporangiophore to a distance said to be sometimes as much as a yard.

Asexual spore types in the Ascomycetes. To most of them the general term 'conidium' is applied, and this name is also often given to those sporangia of the Phycomycetes which germinate directly by germ-tubes; they are, nevertheless, potential sporangia in this group. The conidia of the Ascomycetes may be composed of one or many cells, uni- or plurinuceate, and of the most diverse shapes from simple rounded or oval forms, to branched bodies, sometimes furnished with slender cilium-like antennae from the end cells.

In the simplest case a conidium is merely the end of the fertile hypha or of a small stalk from the end or side of the latter. The spore is cut off by a septum from the parent hypha and may be called a thallospore. Very often, instead of a single spore being thus formed, a chain of spores arises from the sporophore or conidiophore. These chains may be developed in either of two ways; most commonly, each new spore of the chain is successively cut off from the tip of the parent hypha so that the end spore of the chain is the oldest; less often, the first-formed spore buds out another from its tip, and this in its turn another... so that the end spore is always the youngest. As there may be more than "one bud from the tip of a spore, the chance of the second type are often branched.

After a conidium or a chain of conidia is formed on "the end of the conidiophore, the latter may continue to grow from a point just below the tip. The new growth usually tends to preserve the same direction as the original stalk, so that the spore is pushed over laterally. Slight kinks are often visible at the points where such lateral spores have been shed.

Instead of the tip of the conidiophore bearing a single spore or chain of spores, it often branches and the conidia are borne on the ends of the branches, which may be of limited growth and characteristic shape as in Penicillium. Or instead of branching, the apex of the fertile hypha may enlarge into a head or columella as in Eurotium which bears a number of spores or chains of spores, again usually on special short stalks. The term sterigma is again often applied to such a spare-bearing stalk and, as it may branch one or more times before the spores develop, sterigmata of a second or third order may be recognized.

The sterigma, or hyphal branch, which abstracts conidia in succession, thus becoming constricted just below its apes, before cutting off the spore, is often called a phalloid, and the abstracted conidium a philosopher, e.g. Penicillium, Margaritispora, Flagellospora. During this development there is no formation of a cross-wall separating the spore from its phalloid, and a cleavage line arises separating the two when the spore is fully grown.

When, however, a newly developed spore is delimited from its parent hypha by a new transverse wall, it is called an aleuriospore. In some forms it is not certain where the conidiophore and the aleuriospore begins, until the separating cell is differentiated. Again, spores formed on a sterigma or hypha, without reference to the growing point of the hypha, are called radulospores, e.g. Botrytis cinerea.

Some fungi develop their spores in two of these ways, thus the last-unatped type may also produce phialospores, and Acrospeira levis may have aleuriospores and phialospores, but others, such as Penicillium produces, asexually, only phialospores. Thallospores, as above stated, are single spores derived from a mere swelling of the end of a hypha, and are not distinct structures, and while the distinction between radulospores and phialosporesis not always sharp, that between thallospores and phialospores is fundamental. Some spores are formed more or less continuous along the hypha, with no sterile internodes in between as are found in sporoaochia, and are called pionnotes in pionnotal cultures sporulation takes place freely within the medium as well as on its surface, but in sporodochial growth, sporulation is almost entirely confined to the surface.

Outside the three main groups, Phycomycetes, Ascomycetes, and Basidiomycetes, there remains a vast assemblage of fungi known only in the conidial stages. These Imperfect species, the Deuteromycetes or Fungi in perfectly are assumed on good grounds to be mainly asexual stages of Ascomycetes.

Their numbers are slightly reduced from time to time by the discovery of the place of one of them in the life-cycle of some as cigerous species, but it seems probable that in many the perfect stage has been lost and the fungus is perpetuated indefinitely in the couidial condition. They include many parasites and amongst them are represented every type of asexual spore and sporophore mentioned above, the simplest forms being designated Hyphomycetes and the pycnidial forms Coelomycetes. Some hesitate to apply the terms genera and species to such imperfect organisms and prefer to qualify them as 'form genera' and 'form species'.

So long, however, as systematics has not reached the point of indicating true affinity there seems to be little reason for this refinement. Names based on characters that can again be identified are necessary to group organisms together into recognizable units, and a fungus capable of independent existence, perhaps indefinitely, in a conidial stage appears to be fully entitled to such dignity of status as may be implied by the use of a properly employed binomial terminology.

POLYMORPHISM

The occurrence of different types of spore and of spore-bearing structures in the life-history of an individual fungus has been indicated already, Ascomycetes, rusts and Fungi being particularly rich in examples. A rust, for instance, may have spermagonia, acedia, uredosori, teleutosori, and the promycelium in its full development. So also an Ascomycete may have a simple Hyphomycetal spore stage and a development of pycnidia, or spermagonia before asci are produced.

In the life-history of some Hyphomycetes the type of spore found may vary, presumably as a result of age and the action of environmental conditions, to such an extent as to place the fungus in different genera according to the prevalent spore form encountered. Thus, the well-known *Heteropatella antirrihini*, to dispersal, are shown between the cause of 'shot hole' of cultivated snapdragon, may appear as a Pseudodiscosia or a Cercosporella, which is actually only a summer conidial stage of the pycnidial form Heteropatella on the dead or moribund stems: and Ramularia vallisumbrosae the cause of 'white mould' of narcissus, may appear as an *Ovularia* or a *Cercosporella* instead of a typical *Ramularia*.

Similarly there is reason to believe that *Ovularia primulae* and *Cercosporella primulae* are the same fungus as *Ramularia primulae*. Amongst pycnidial fungi, again, it is probable that Phoma destructiva and *Diplodina lycopersici* refer to the same tomato parasite, the perfect stage of which is *Didymella lycopersici,* while it seems likely that *Ascochyta lycopersici* is only another phase of this polymorphic species. Species of *Septoria* and *Stagonospora* are also stated at times to pass through an *Ascochyta* stage. It is cases such as these that give force to the arguments against recognising imperfect forms as unqualified genera and species, but many of them are due to such notoriously variable characters as spore septation and sporophore type. Indeed, it is not unknown for variable characters in the sporophore to make it difficult to assign, say, one of the *Polyporac*eae to its proper genus.

In the opposite direction to the conidial forms mentioned in the last paragraph are those in which similar conidial stages may occur in fungi that are systematically far removed from one another. Thus both the Ascomycete.

THE DISSEMINATION OF FUNGI

A few fungi are disseminated principally in the mycelial state as, for instance, the tropical thread blights which blow about with the leaves to which they adhere; the 'Corticium' disease or 'red thread', a common trouble of lawns and sports greens in Britain, produces on the leaves a mycelium of coral-like threads which adhere to the tips of the dying leaves. Others are distributed as vegetative propagants such as the sclerotia and stromatic crusts of Rhizoctonia solani, of which the sporing form, *Corticium solani,* plays little part in spreading the fungus. The long-distance dissemination of fungi is, no doubt, often to be traced to the transference of viable mycelium in plant produce from one country to another; this has happened in elm disease and very probably in potato blight and with many other fungi.

Usually, however, the spores are the organs of dissemination. As they are very small and so light as easily to be caught up into the air as an impalable dust, the wind is the chief agent in their dispersal; there are, however, other important methods of spread.

The spores of fungi may retain their vitality for considerable periods and they show amazing powers of resistance to extremes of heat and cold: sealed in vacuum tubes, slowly dried spores of Mucor and Aspergillus have withstood a temperature of liquid air for nearly 500 hours, followed by immersion of the tubes in liquid hydrogen for 77 hours. Susceptibility to heat varies greatly, for some spores are killed in 10 to 15 minutes at temperatures far below the boiling point, whereas occasionally some can stand the direct action of boiling water or live steam for the same length of time. So also the spores of some species are killed by drying for 24 hours or even less, while others can be best preserved in a dry condition.

As a rule spores are less sensitive than hyphae to adverse surrounding conditions, and vegetative growth is often inhibited at, or near, the freezing point, as is evident from the action of refrigeration in checking mould growth; slow growth on frozen foodstuffs, however, has been recorded, and very slight superficial moldering of meat will occur at a temperature of – 6° C. Unless special percutaneous are taken, moldering as a result of spore germination quickly supervenes on removal from cold storage. These resistant properties of spores are of value in facilitating the aerial dissemination of many fungi, for desiccation in the air and extremes of beat and cold on the ground may have to be faced.

The vertical elongation of the sporophore is often a powerful aid to exposing the spores freely to air currents. Amongst other adaptations which seem to have the same object, the forcible projection of spores from their stalks is often important. When enclosed in a receptacle there are sometimes arrangements for expelling the spores actively into the air. These are often brought into action by changes in temperature or dryness, as in some Ascomycetes where' puffing' can be induced by concentrating the sun's rays through a pocket~ lens on the.receptacle. Some of the Discomycetes respond similarly if breathed upon when ripe. In the Basidiomycetes it is usual to find a mechanism for the violent release of the basidiospores based on the secretion of a drop of fluid at the junction of the spore and its sterigma. For full advantage to be gained from these various adaptations it is necessary that the spores should be dry and free from mucilaginous envelopment.

In a considerable number of species, including many fungi whose spores are developed in acervuli, the spore beds are surrounded by a gummy secretion. A great many of the bacterial plant pathogens also are immersed in mucilage as they are extruded to the surface of the host plant. When dry, the secretion tends to harden into flakes or sheets, but these are readily soluble in water, as in 'halo blight' of French beans, due to a bacterium.

Dissemination may be effected by the blowing about of the dried gummy shreds, followed by the setting free of the individual cells in rain drops, as sometimes occurs in the American fire blight of fruit trees, due to Bacterium amylovorum, or in *Xanthomonas malvacearum* in cotton, though the latter is often also carried as dust from adjacent cotton fields infected the previous season. The spore bed may be exposed to rain splashing which carries the spores to leaves or stems in the vicinity; species of *Colletotrichum* are often disseminated in this way. Amongst the Hyphomycetes it is possible to distinguish two main groups according to the slimy or dry character of the spores, and the method of their dispersal depends largely on this character, the dry spores being predominantly suited for distribution by the wind, while the slimy ones are carried by water, by insects, or by adhering to adjacent plants and the like. In a recent survey of the British Hyphomycetes three-fourths of the species belonged to the dry-spored group. The Coelomycetes, on the other hand, are generally representative of moist sporing fungi. In some of them and in many of the moist-spored Hyphomycetes, the spores are liberated in a mass held together by water or mucilage, and in this form may be subject to aerial dispersal before they are broken up by rain or dew.

Knowledge of the long-distance air-borne dissemination of fungal spores has greatly increased of recent years.

Observations on the spread of white pine blister rust in the west of the United States and Canada indicate that it is due to wind-blown infection of currant and gooseberry bushes by aecidiospores from pines that may be hundreds of miles away. Similarly, cereals are infected over great areas by wind-borne uredospores of their rusts. Numerous examinations of the spore-content of the upper air have been made by aeroplane. In early flights, carried out in 1921, spores and pollen grains were found to be fairly abundant up to 11,000 feet, above which they became scarce.

U redospores of *Puccinia graminis* caught at 7,000 feet were germinated successfully. Above 10,000 feet, mould spores, such as those of Aspergillus and *Penicillium,* and bacteria, are most frequently found, occurring at times up to nearly 20,000 feet. There is circumstantial evidence that the distance to which such spores may be borne in a viable condition during storms is limited mainly by their ability to survive atmospheric conditions. These naturally vary greatly from year to year during the period of spread, but it is reported from Canada that there is rather conclusive evidence that under certain conditions the black rust of wheat has been carried for four or five hundred miles, or even farther, through the air. These spores tend to be deposited in definite showers rather than in a continuous steady fall; in western Canada rust spore-showers occur after a day or two of strong south wind.

The long-distance disstfmination of cereal rusts has been intensively studied. Black rust due to the well-known parasite Puccinia graminis occupies a practically continuous wheat belt in North America from the Gulf of Mexico to the Prairie Provinces in Canada; as the crop is maturing in the south at the same time that it is being sown in Canada, and the intervening areas are more or less intermediate, a regular succession of wheat ready for infection is available from south to north in the spring and summer. The spores seem to follow a main northward drift path from an area in Texas, where the rust over-winters in the uredo stage and the wheat ripens in the late spring, through Kansas in May and June, to the upper Mississippi States. In Texas itself the summer is intensely hot and the rust is unable to survive until the new crop is sown in the autumn. Reinfection then takes place, apparently chiefly by a reverse flow of spores from the north. So there is an ebb and flow of infection to and from the area in which this fungus over-winters but cannot survive the summer. Confirmatory evidence of this movement is provided by the incidence in particular years of certain physiologic races of the fungus in the areas concerned.

No doubt the spread is not ordinarily carried out in great jumps, though spores not of local origin have been caught as far north as Nebraska in May and June: usually there is time for one or more breaks and the renewal of the inoculum by the infection of local crops in the path from Texas to the northern Mississippi and Canada. Some irregularity of incidence in the upper part of this tract may result from the local over-wintering of the fungus in the teleutospore stage and infection of the alternate host, the barberry, in the spring, for this rust is heteroecious and produces its aecidial stage on barberries, whence it can infect neighbouring wheat and possibly start an epidemic.

But many of the barberries have been destroyed in the barberry eradication campaign and their disappearance in due course will, it is hoped, save much of the chief wheat-growing region in the northern States and Canada from all but the northward spore drift, for the region is much too cold to allow the rust to over-winter in the uredo stage. Attention can then be concentrated on the production of rust-resistant wheats in the comparatively restricted over-wintering area in the south. Nearly all the wheat grown in India is around as a winter crop in the plains.

In these areas rust is unable to survive the intense heat of the summer, and the wheat and barley crops are infected anew each year from rusted cereals grown in the Himalayan foot-hills and some of the hills of Peninsular India. Barberries occur in these hills but.not in the plains, and though the aecidial stage of Puccinia graminis has been found on several of the Himalayan species, it seems to play little part in the perpetuation of the rust. The uredo stage is abundant when the hill crops are harvested in May and June, and then persists on stubble shoots and volunteer plants until the new hill crops are sown in the autumn. Areas have been found in the Central Himalayas where the early crops are sown in August-September and may be heavily rusted by the first week of December. Inoculum is thus available in time to account for the first outbreaks in the Indo-Gangetic Plain, which usually occur in January-February.

The uredospores have been trapped in the air by means of specially devised aeroscope slides we use before the rust has been found on the crops in the plains. In Peninsular India similar foci of early infection of the growing hill crop have been found and seem to account for extensive distribution of the rust in Madras and the Deccan. A similar chain of infection serves to explain the distribution of wheat brown rust whereas Puccinia glumarum, the cause of yellow rust, is more sensitive to heat and its uredosori are not ordinarily found in the summer below about the 6,000 feet level in the Himalayas, though it may be the only rust seen above 7,000 feet in N ovember- December.

The long-distance dissemination of cereal rusts in several parts of the world is effected on the same broad lines. Thus the wheat in the Amur region of eastern Siberia gets its inoculum from over-wintering areas in Manchuria; North Africa seems to be infected from Europe, the Balkans from Asia Minor, and the Argentine from Brazil. So to a great extent has been cleared up one of the greatest mysteries with which plant pathologists in certain parts of the world were faced before it was known that spores could be carried over long distances in the upper air.

Such long-distance spore flights are evidently limited to certain types of spores and of prevailing meteorological conditions, at least so far as the parasitic fungi are concerned. If they were general, the object of the plant quarantine services maintained by most countries would be defeated. Heavy wind-drifts or storms no doubt play an important part in them, as does the presence in a receptive condition of a susceptible crop in the area reached. It is known, furthermore, that sporecontaminated air not far from the ground purifies itself by the fall of the spores at rates varying from o. 5 to 5 mm. per second, while their presence over the ocean has been found to diminish with increasing distance from the land and catches taken on shipboard may fall to nil in individual voyages fairly soon after reaching the open sea; various favourable conditions must coincide to permit successful long-range spread, and they do not seem often to occur.

There is still a great weight of evidence from the known facts of the spread of parasitic fungi into new areas in support of the view that' discontinuous' dissemination is usually effected by transport on or in the host plant, and is linked with the movement of nursery stock or other plant produce along recognised trade routes. The discovery of spore-passage in the upper air and its bearing on the dissemination of the rusts shows that wind-borne distribution can be an important factor in certain circumstances and must always be regarded as a possible menace by those responsible for the protection of crops against foreign disease.

It is also ordinarily accepted as responsible for much of the' continuous' dispersal of fungi within any given land area. But sea-barriers and sometimes mountain ranges have proved effective checks to the dissemination of many plant parasites. Either the spores are unable to survive long passages in the air, or the timing of their flight is such that a suitable host in a receptive condition is not available when the new area is reached, or the meteorological conditions are adverse, or other causes intervene to spoil a safe lodgement.

Many favourable circumstances must combine to render the invasion successful, and experience suggests strongly that they seldom do. Supporting evidence can be found in the history of the spread of American gooseberry mildew in the British Isles and on the continent of Europe; of potato wart disease in many parts of the world, though the resting sporangia of this fungus might be expected to reach the air with dust from infected fields; of blister blight on tea in India; of cacao witches' broom in South America and the West Indies; of several parasites of trees and of ornamental plants, and of many others. In general it may be said that if human agency be excluded, dispersal within a new area is gradual and continuous, each new generation extending the area by a rather limited distance.

This happened when the, vine mildews and that of the oak were introduced into Europe from America, though the latter at least seems to have had some jumps of about 100 miles, and also with several diseases such as chestnut blight introduced into America from abroad.

Some spores are disseminated largely with the seeds of the plants on which they grow. This is the chief means of spreading bunt of wheat and various other diseases against which seed disinfection is practised. Soil fungi are carried by irrigation water or surface-wash after heavy rain and are also liable to be distributed by farm implements, or plough cattle, or on the feet of farm workers.

The dissemination of the spores of certain fungi by insects has been noticed by botanists for very many years. Basidiomycetes belonging to the Phalloid or stinkhorn group were particularly marked examples, for their strong and penetrating odour attracts carrion and other flies, and the special adaptation of this group for insect dissemination was early suggested and subsequently fully established. The sweet honey-dew in which the spores of some fungi are extruded has also long been known to attract insects. These visit the 'Sphacelia' stage of Claviceps purpurea, the cause of ergot of rye, and carry spores away not only as an external contamination but also in the alimentary tract, through which they can pass uninjured. The more recently discovered function of insects in securing the fertilizations of the rusts has been mentioned above.

Naturally, however, it is the role of insects in the spreading of parasitic diseases that has attracted most attention in modern times, and its importance is now fully recognised. Insects are the plain agents of dissemination of virus diseases. They also play a considerable part in the spread of bacterial diseases, such as fireblight in orchards, where they are attracted to the gummy exudate from twig cankers in the spring and carry the bacterium-laden matter to the blossoms and young growth of healthy apple and pear trees; the honey-bee is considered to be one of the chief agents in initiating the blossom blight form of infection, and the discovery that the organism is able at times to persist in the hive was followed by an embargo being placed on honey from infected countries in some places still free from the disease.

Some of the insects concerned are able to inoculate healthy trees by puncturing the young growth. The organism that causes bacterial wilt or Stewart's disease of maize,*Xanthomonas stewarti*, has been found to over-winter in hibernating *Chaetocnema pulicaria* flea beetles, the chief vectors of the disease in the United States, and to be inoculated into the growing crop by these insects in the spring. Cotton stainers of the genus Dysdercus, and some other insects, are responsible for the introduction of the minute fungi *Nematospora* and its allies into young cotton bools, where they induce internal boll disease; the insects live in between successive cotton crops on various plants and trees, of which the baobab is one of the most important in Africa, but it is not yet known how or where they harbour the fungus, though the evidence suggests that the association is close.

In many cases insects that cause wounds, such as those that bore into stems or gnaw the softer parts, carry parasitic bacteria or fungi on their bodies and serve to spread infection. The Dutch elm disease is mainly disseminated by bark-boring beetles of the genus *Scolytus,* and the watermark disease of willows seems to be spread by several different kinds of insects.

The black rot, and soft rot of cruciferous plants, due to Xanthomonas campestris and Bacterium carotovorum respectively, are also carried passively by insects and other small animals such as slugs. It is easy to demonstrate how readily some of these organisms are picked up by allowing an insect from a heavily diseased plant to pass across the solid surface of some suitable nutrient preparation and observing the colonies of micro-organisms that subsequently develop along the insect's track.

THE NUTRITION OF FUNGI

The spare, on germination, gives rise to a germ-tube, which at first is nourished by reserve protein, carbohydrate, and fat stared up in advance. In many cases the fat reserves are mare prominent than those of a protein or carbohydrate nature, being often very noticeable as oil drops in the ripe spore. Very soon the young hyphae begin to feed far themselves. The organic food they require must be dissolved before it can be taken in, because solid particles cannot pass through the cell membranes.

This food in solution reaches the interior of the hyphae by a process of absorption. Unlike the green plants, fungi cannot build up their carbohydrate intends by photosynthesis from the carbon dioxide of the air, far they lack the chlorophyll which makes this possible. In this respect the chlorophyll-free plants are like animals requiring their carbohydrates to be ready prepared by having been built up to farm the bodies of other plants or animals. They are usually capable, however, of modifying carbohydrates and other organic compounds by enzyme action sa as to make them mare readily assimilable.

In organic substances-mineral salts and elements like nitrogen are as necessary in the nutrition of fungi as in that.of higher plants and animals. Mast.of these, however, occur in combination in the.organic substances.on which fungi feed. As with the higher plants nitrogen, phosphorus, and potassium are amongst the mast important of these elements.others required in lesser degree being sulphur, magnesium, calcium, and a little iron.

It is nowadays a matter of common knowledge that animals and man require certain accessory food substances vitamins and the like in addition to those which satisfy their needs in proteins carbohydrates, fats, and salts. Fungi seem to have similar requirements, though knowledge of these is not yet far advanced. Many fungi and bacteria are able themselves to synthesise these substances and even to produce them in excess and excrete them into the surrounding medium. Various examples of the growth or development promoting action of one fungus an another have been known far a long time, and mast of these can be traced to the liberation of auxins or vitamins from the mycelium of the first of the two organisms concerned. Filtrates of the staled media in which certain bacteria and fungi have been cultivated have the same effect.

Aneurin has been found especially active in this respect. Many fungi are able to manufacture sufficient far their needs or even to produce an excess, but others require an external source of supply. On the whale, this part of the vitamin B complex appears to be the mast generally valuable accessory substance required by fungi.

Yeasts grown in pure synthetic media develop slowly until they synthesise a substance originally named 'bias' but now known in purified farm and commercially available under the name 'biotin' when biotin or various fungal or bacterial filtrates or organic substances containing it are added, growth is considerably stimulated. A good many other fungi are now known to respand by increased growth when biotin is made available to them; same samples of the agar used far solidifying culture media contain it in considerable quantity.

Sometimes two fungi can be grown successfully in mixed culture in synthetic media when either alone makes little progress. Thus, Polyporusadustus and Nematosporagossypii do well together, the former supplying biotin which is essential to *N. gossypii* and the latter aneurin without which *P. adustus* virtually fails to grow. Similarly the red yeast Rhodotorula rubra can grow together with Mucor ramannianus in a synthetic medium in which neither alone succeeds. Both biotin and aneurin can be synthesised by certain fungi from pure dextrose, amino-acids and inorganic salts.

According to the manner in which fungi obtain their organic food, they are divided into two great classes: Saprophytes and Parasites. Saprophytes obtain their organic food from the dead tissues of animals or plants, or of substances derived from them. Hence they are found in large numbers in the soil, pervading every bit of rotten leaf or twig or root or every particle of manure. They grow as moulds on foodstuffs, and in the tropics on boots and the backs of books in the rainy season. Every bit of old timber in the forests is liable to be permeated all through with their hyphae. As already mentioned, they share with the bacteria the role of nature's scavengers, often having a specified part to play in preparing refractory material, such as the skeletal parts of plants, for bacterial decomposition. In breaking down the walls and contents of plant cells, they liberate carbon dioxide which becomes available in the air for the carbohydrate nutrition of green plants, hydrogen which forms water, ammonia, and so on.

Parasites obtain their food from the living tissues of animals and plants. While, therefore, many saprophytic fungi playa useful part in nature as scavengers, the parasites are almost entirely hurtful. The few cases of beneficial parasites, from the point of view of man, are those which attack noxious insects and occasionally destroy large numbers of them. The great majority of parasitic fungi feed on living plants, those useful to man as much as, or more than, the weeds or useless kinds. A living plant on which a fungus feeds is often termed a 'host plant' for that fungus. There is scarcely a plant of our fields, gardens, and forests which does not serve as a host plant for one or more species of fungi.

Several different classes of parasites may be distinguished. Obligate parasites, for instance, pass through the whole of their life-history on living plants, and cannot be grown on dead or artificial food material. Such are the rusts, the powdery mildews, some Ascomycetes, and other fungi. Facultative saprophytes are those that feed usually on living tissues but can, at need, pass through a part of their lives as saprophytes. Often their full development is only reached on the living plantsaild they cannot be got by artificial cultivation to complete the whole of their life-cycle. The smuts furnish examples of this, as they can often live by budding out little conidia for a considerable time in soil, but usually only grow to maturity and produce their perfect spores on a living plant. Facultative parasites, on the other hand, are those which usually grow on dead or decaying matter and are capable of passing through the whole of their development as saprophytes, but have at the same time the faculty of attacking living tissues under certain conditions.

Within these main divisions, parasitic fungi may show every degree of adaptation to the parasitic mode of nutrition. At the lowest end of the scale are certain facultative parasites, which may be called' weak parasites', as they become parasitic only occasionally, when the host plant has been weakened in its vitality by some harmful agency. Plants not yet acclimatized in a new locality, or suffering from insect attack or mineral deficiency, or grown in insanitary conditions such as overcrowding, too dense shade, unsuitable or badly drained soil, and the like, are particularly liable to become available as food to these fungi that are normally saprophytic.

Cases of the sort have been recorded from time to time, though few have been adequately investigated; for instance, Diapor the perniciosa and other fungi to which the dieback and bark canker of fruit trees in the south of England were at one time attributed are now believed to attack only trees weakened from other causes. It is, perhaps, as profitless as it is impossible to draw a hardand-fast line between saprophytism and parasitism in some of the fungi; even in the commonest of all species, Cladosporium herbarum on fading green plant parts, there are too many records of injury from its action to deny it all possibility of behaving as a weak parasite under certain conditions.

In many cases, particular tissues of a plant are normally of low vitality and are readily available as food for weak parasites. Such are the gorged cells of many pulpy fruits, oranges, cherries, apples, and the like, and also the older living wood of trees.

It is probable that the fungi termed 'wound parasites' are limited in their aggressiveness by some nutritional failing. At first these live only on the dead and rotting tissues caused by a wound or injury to the plant. Later on they begin to spread to the undamaged cells farther in, and they may extend through a large area of the living tissues, eventually, perhaps, killing the whole plant. It may be that during the period of saprophytic life the enzymes of the fungus are produced in quantity and concentration sufficient to enable them to attack living tissues exposed to their action, and transform the walls and cell-contents into nutrient material which the fungus can utilise. If this is so, an analogy can be drawn with the action of decaying petals in promoting infection by Botrytis cinerea. That enzyme production can be greatly modified by nutrition has been fully established in this and other fungi.

The distinction between parasites and saprophytes cannot be maintained in a few fungi which are able to attack perfectly sound tissues of healthy plants should they come in contact with a suitable host, but in the absence of the latter can live and develop freely on dead materials. Such a case is the' damping off fungus *Pythium de baryanum* which is present as a saprophyte in almost every garden in some parts of Europe, and at once attacks and destroys seedlings of cress and certain other plants, should they be grown in the soil which contains the fungus.

Neither neglect of cultivation nor unsuitability of soil predisposes to this disease; wounds are not necessary for the entry of the parasite; and, if anything, damping-off is worse on rich soil than elsewhere. Several parasites of trees, e.g. the common bracket fungus on elms, *Polyporus squamosus* can continue to live on the dead wood as saprophytes; these, however, are usually wound parasites of a less aggressive type than the damping-off fungi. On the Aleppopine in France, *Trametes pini* dies with the tree, but *Fomes pinicola* continues to rot the fallen timber.

THE MECHANISM OF FEEDING BY PARASITES

Parasites reach their food either by sending their mycelium into the interior of the host plant or by remaining for the most part superficial and sending only suckers or a limited growth of feeding hyphae into the surface tissues. Usually in the internal parasites, e.g. *Rhytisma acerinum,* the feeding hyphae penetrate directly into the cells of the host, the cell sap and more solid contents being taken up by absorption after the latter have been brought into solution by the enzymes of the fungus. Often, however, the earlier stages of feeding are marked by the dissolution of the cell walls of the host by enzymes such as pectinase and cellulase; these walls can provide many fungi with much of the carbohydrate food that they require. In a few fungi, such as the *Exoascaceae,* the mycelium is wholly intercellular, without any intracellular parts, so that absorption of nutrient solutes must occur through the membranes of the host and parasite cells; this is probably a common occurrence, for even when the cells are entered much of the mycelium may remain intercellular and probably receives its food directly from the surrounding cells. It was at one time believed that some of the superficial parasites of leaves and twigs could similarly feed by absorption from the epidermal cells across the outer wall of the cell and the cuticle, but in most of these fungi, feeding hyphae or suckers have now been found to enter the cell 'cavities; this occurs, for instance, in the light-coloured thread blights and in the dark *Meliolaceae* or *Asterineae*.

In the ectoparasite of some European pine trees, *Acanthostigma parasiticum,* it is stated that the white mycelium which forms cushion-like layers on the under surface of the needles obtains its food through tiny rod-shaped processes which penetrate through only the cuticle and not through the inner parts of the outer wall of the epidermal cells. As in the subcuticular parasites such as *Taphrina rhomboidalis,* absorption of nutrient solutes must occur through the subcuticular membrane of the needle leaf.

HAUSTORIA

Many intercellular parasites that apparently find difficulty in getting adequate nutrition by absorption of solutes into their hyphae from the cavities of the host cells send into the cells specialised hyphae which act as feeding organs. These are known as 'haustoria'. Haustoria occur in all the main groups of fungi, usually but not exclusively in obligate parasites with an ectoparasitic or an intercellular mycelium. They vary in shape from the knob or saclike type found in Cystopus to simple or lobed or branched or coiled hyphae in many *Erysiphaceae* Phycomycetes and rusts, or coralloid types as in *Peronospora schachtii* on beet. Some of the complex forms almost fill the cells they occupy and can hardly be distinguished from an intracellular mycelium, as in *Peronospora parasitica*. Their shape is not very constant, as it may vary to some extent according to the tissue invaded.

The haustorium usually enters the cell by an exceedingly fine hypha resembling the infection hypha arising from an appressorium by means of which, as will be seen later, many parasites enter the outer membrane of the host. After entry the haustorium often grows towards the nucleus of the host cell but while, at first, it may not enter the' primordial utricle' or peripheral plasmatic membrane of the protoplast, only pushing it in and invaginating it, the utricle may finally be pierced by it. In the early stages the haustorium is bounded by a thin limiting membrane, and this sometimes persists. Frequently, however, various thickenings occur around the point of entry or deeper in.

When developed around the narrow style-like stalk of the haustoria of *Erysiphe graminis*, or many rusts, the thickening is continuous with the inner wall of the cell and is closely similar to the infection pegs found where a parasitic hypha enters a cell. It is sometimes effective in checking penetration. When found around the body of the haustorium it is often later in appearing and may, in old haustoria, be a means of walling off the intruder, or a consequence of excessive drain on the cell contents, as in tubers attacked by potato blight; as injury to the cell from over-virulent action of the fungus increases, so the haustorial cover thickens.

There is little doubt that these thickenings are products of the host cell, and indeed it is sometimes possible to distinguish two layers in the sheath, one being the membrane of the haustorium and the other a covering deposited by the host. The latter seems to be usually cellulose, but there may be some lignification of the older sheaths. Callose is stated to occur sometimes between the two layers or in the exterior one. The haustorium itself may have from one to many nuclei. In some smuts the haustoria are typically binucleate, but those of the downy mildews may sometimes have only one nucleus though others have several, as in the coenocytic hyphae from which they arise.

Most of the haustorial fungi arc obligate parasites, but some, such as the Phytophthoras and *Actinonema,* can be grown in artificial culture. It is probable that many of the intercellular hyphae are not wholly dependent for their nutrition on the haustorial apparatus, but, like other fungi occupying a similar position, can get some nutrition by absorption from the surrounding cells.

While, therefore, the haustorial habit is not a strict criterion of status as an obligate parasite, it suggests a high degree of specialisation in some requirement which ordinarily it has not proved possible to supply in culture. It marks increased adaptation to the parasitic life over that shown by the destructive action of parasites such as *Pythium* and *Botrytis,* for one of the characteristic features of many of the haustorial fungi, especially the rusts and smuts, is the absence of any gross indication of injury to the protoplast of the host cell for a period which may sometimes be almost equal to the lifetime of the plant. This aspect of their parasitism will be referred to again in a later chapter.

THE FUNGI ENZYMES

In the preparation of their food for absorption fungi, like animals and man, rely on enzymes or ferments, of which they may possess a formidable array. Among the hydrolytic enzymes detected in Polystictus abietinus are cellulase, pectinase, ligninase, diastase, sucrase, inulase, tanninase, pepsin, trypsin, emulsin, and lipase, while the oxidising enzymes of this fungus include catalase, peroxidase, and laccase. In *Aspergillus oryzae* diastase, invertase, maltase, lactase, protease, rennet, and lecithinase have also been recorded.

Many of these enzymes are excreted into the surrounding medium, sometimes in considerable quantity and capable of utilisation on a commercial scale. Apart from the use of yeasts in brewing, distilling and baking, diastase from fungi is used medicinally and in various industrial processes; acetone can be formed by *Mucor rouxii* and other fungi from sugar-yielding plant products; citric acid is now mainly a fermentation product of *Aspergillus niger*, for which one American firm some years ago kept nine acres of mycelium in constant commission; yeasts, Mucors, Aspergilli, and bacteria are all employed singly or in various combinations in the East and in Africa for the preparation of various 'beers', fermented milks, non-alcoholic drinks, and seasonings; while gallic acid, which is used in dyeing and as a constituent of inks, has been commercially produced by the fermentation of tannins. The proteolytic enzymes of *Aspergillus oryzae* and its allies are used in the Far East in the preparation of readily digestible substances from the protein-rich soy bean.

Not less important are the processes of fermentation carried out during metabolism within the hyphae. During the period of stringency of fats in Germany about 1918 more than a million kilograms of glycerine were being produced monthly from fungi, and in Russia *Endomycopsis vernalis* was used experimentally for the same purpose some years later. Molasses, beer-wort, potatoes, hydrolysed wood pulp, or the like, together with salts of nitrogen and phosphorus, provide the culture medium for the fungus. Some moulds are even more effective in synthesising fat as a reserve nutrient in their mycelium than the yeast-like *Endomycopsis;* Penicillium javanicum can store up to 41.5 per cent of fat in its hyphae when grown on glucose and salts.

Yeasts and their allies are also able to manufacture high-quality proteins, with vitamins, from the same raw materials, and some 20,000 tons of dried yeast, containing about half its weight of protein, were annually produced in Germany during the first great war. Probably the best species for this purpose hitherto tested is Torula utilize. These examples, taken from only a few amongst many, indicate that both by means of the enzymes excreted from the mycelium and by those concerned in metabolic processes within the hyphae fungi can synthesise a remarkably wide range of substances from organic materials. Many have been identified and they represent almost every type of compound known to organic chemistry.

The enzymes are organic catalysts which, unlike the inorganic ones, are destroyed by heating. Hydrolysing enzymes split up disaccharides such as sucrose, maltose, lactose, and the like into two similar or dissimilar molecules of monosaccharides-dextrose, levulose, mannose, galactose etc. Cane sugar, for instance, is readily' inverted' by invertase from fungi giving one molecule of glucose and one of levulose, while maltase gives two molecules of glucose from maltose. The hexose monosaccharides, especially glucose, appear to be readily available to fungi; on condensation they give hexosan polysaccharides.

The pentose monosaccharides, such as xylose and arabinose, condense with other sugars to form pentosans-polysaccharides common in pectic membranes, gums, mucilages, and the like; acted on by the pectinase of fungi, they give galactose and other bodies. On splitting up, the polysaccharides give more than two molecules of monosaccharides. The proteolytic enzymes of fungi, pepsin, trypsin, erepsin, etc., change the less soluble proteins into the more soluble and readily diffusible peptones and amino acids. Tannase forms glucose from tannin and lipase, glycerol and higher fatty acids from fats. Certain moulds can grow in the waste liquor from the manufacture of sulphate wood pulp, decomposing the lignin present as chemically combined lignosulphonates.

The oxidative processes of certain bacteria provide energy enabling them to synthesise carbohydrates when carbon dioxide is their sole source of carbon, although they have no chlorophyll and light is not needed for the purpose. Such are the nitrifying bacteria, Nitrosomonas and Nitrococcus, the sulphur bacteria. *Beggiatoa* and *Thiothrix*, and the iron bacillus Spirophyllum. The oxidase of strains of the common mould *Aspergillus niger*, can oxidise glucose to gluconic acid directly. Similarly some yeasts can directly oxidise maltose, giving off carbon dioxide. Yeasts will grow with the organic salt ammonium tartrate as their only source of carbon. Such examples are sufficient to illustrate the remarkable versatility of the enzymes of bacteria and fungi.

The enzyme armament varies considerably in different fungi. Some can attack chiefly the cell membranes, others the cell contents. The *Peronosporaceae, Erysiphaceae,* and rusts, for instance, have little or no action on the membranes, using chiefly some of the cell contents, while those wood-decaying fungi that affect tissues with scanty cell contents must get much of their food from the walls; this accounts for the cavities in the older wood caused by certain species. Enzyme production may vary according to the strain of the fungus concerned, as is fully recognised in brewing and other fermentation industries.

The conditions under which the fungus grows, especially its nutrition and aeration, also have a marked effect. Thus a strain of Aspergillus niger was found to form diastase copiously when grown in media containing starch, glucose, or maltose but not at all when grown on 5 per cent glycerine. An acid reaction was necessary for optimum production on the enzyme in this case, but pectinase may sometimes be most active in acid, sometimes in alkaline plant juices, and this seems to depend on the presence of other substances which are adsorbed from the medium.

It is believed that the pectinase is the same in all of a number of fungi tested, but its properties depend on its exposure to the action of other substances, while whether it is produced at all and the quantity produced, depend on the medium in which the fungus is grown.

Some fungi can cause notable rises in temperature in decomposing plant material. Thus Aspergillus fumigatus has been found to raise the temperature of oat straw inoculated with it from 25° to nearly 55° C., in 38 hours. The most extensive decomposition of straw has been obtained by a mixture of soil Actinomycetes and fungi. Most of the common soil moulds Aspergillus, Penicillium, Trichoderma, Rhizopus, and so forth utilise the carbohydrates cellulose, hemicellulose, pentosans, and mannan readily and rapidly in decomposing plant material but not lignin; certain Basidiomycetes! however, such as the common edible mushroom, do not make use of hemicelluloses but deplete the lignin and protein, while causing also a loss of cellulose. Some of these fungi can decompose humus.

NUTRITION AND LIFE-HISTORY OF FUNGI

All saprophytes and many parasites can be grown on artificially prepared food, or 'culture media'. The kind of food supplied may have a profound effect on the structure and reproduction of the fungus. With such plastic organisms, however, generalisations are difficult and the most that can be said is that a particular effect is obtained with a given strain of the fungus grown under a certain set of conditions. Other strains and other conditions may give different results.

In some fungi, as in higher plants, it is possible to enhance vegetative vigour at the expense of reproductive activity by nutritional treatment. Thus, the aquatic fungus Saprolegnia mixta tends to produce a vegetative mycelium only, without any spores, when supplied with abundant food regularly renewed. If such sterile, strongly growing cultures be transferred to water, or to a weaker food solution, sporangia with zoospores are formed in quantity.

Transferred to a solid substance of little nutritive value, such as plain agar, sexual oospores are alone produced. In early studies of *Pythium ultimum* only oospores were produced on cabbage leaves whereas on other substances sporangia were formed. The mycelium of various common moulds is quite different in structure with different food supplies; in some of the Mucors it may be changed wholly or in part from the filamentous to the gemmate condition. Nutrition may determine the proportion of aerial to submerged mycelium or the formation of coremia in place of single hyphae, or even of sporodochial erect clustered hyphae instead of a recumbent growth.

In some genera the colour of the mycelium may vary according to the composition of the medium on which it grows and also according to the pH of the latter. Such effects are the commonplace of all mycological laboratories and account for the caution shown in attempting to identify fungi from mycelial characters alone. As already mentioned, the morphology of the spores and their method of attachment to the sporophore are much more reliable specific characters and are those mostly used. Even here, however, there are difficulties, linked largely with vigour or luxuriance of development. To surmount them, many futile attempts have been made to devise a standard or an ideal culture medium on which fungi would give consistently satisfactory growths.

That these must fail in, at least, the considerable number of facultative parasites which show the specialisation of parasitism described in the next chapter is evident, if specialisation is dependent on nutritional qualities of the host, and there is overwhelming experience available to show that even pure saprophytes do not all use the same food and cannot all be given a single satisfactory diet.

SYMBIOSIS

The parasites mentioned so far obtain their food at the expense of the host plant and are not known to give anything in return. The food-supplying cells are usually killed eventually by the excessive demands of the invader, but, as already mentioned, certain obligate parasites, especially those that feed by haustoria and those that form galls, may long defer this critical period and kill only when the production of spores makes a culminating call on the nutritional resources of the host. In many cereal and grass smuts this does not happen until the ear or individual caryopses are developing at the end of the vegetative life of the host. The Ascomycete *Epichloe typhina* may go to a stage further than this, for it sometimes fails to fructify even when its systemically infected grass host has grown to maturity.

The mycelium may be found in the anthers and around the embryo of the grass seed without necessarily destroying the pollen or preventing the seed from germinating. Latent infection may persist for several seasons, the plants sometimes remaining barren, sometimes producing viable seed. Ultimately the conidial stromata appear during a flowering period, which may be in the second or third year of growth; up to this time no apparent injury is suffered by the host. Perhaps the end point in this progress towards a regulated parasitism is marked by the endophytes which occur at times in various species of Lolium. Two of these have been found to cause systemic infection of the above-ground parts of the hosts, one difficult to isolate and remaining sterile in artificial media while the second grows readily in culture and produces microconidia: the former develops a dense mycelial stratum between the seed coat and the aleurone layer, the latter is sparser and more difficult to detect. Both are transmitted in the seed, from which they pass to the growing point of the seedling.

Neither appears to cause any injury, and their presence in or absence from a particular plant cannot be detected without isolation or microscopical methods. Their identity is unknown, and there is no real clue to the physiological relationships between the host and the parasite, though there is some experimental evidence which has been considered to indicate that the endophyte assists in the supply of nitrogen to the host by fixing the nitrogen of the air.

So far all the advantages seem to lie with the parasites; the host at best may be able to escape injury to a greater or lesser degree, or perhaps, in Lolium, gain a little nitrogenous food by the aid of the fungus. Some higher plants, however, have progressed beyond this stage and have established a symbiotic partnership with fungi in which a mutualistic relationship is evident. As the association between the fungus and higher plant is chiefly evident in roots, the term 'mycorrhiza' is applied to it.

The best-known mycorrhizas are those of forest trees and orchids: these represent the two main types of fungus-root association, the 'ectotrophic' and the 'endotrophic'. In the ectotrophic mycorrhizas of trees many of the fine lateral rootlets are enveloped in a mantle of fungal tissue which prevents growth in length of the rootlet, causing it to assume a swollen or coralloid form. Some of the hyphae usually pass down between the cells of the outer layers of the root, forming the so-called Hartig net.

The fungi concerned chiefly belong to the Basidiomycetes and include many gill and pore bearing toadstools but truffles grow in mycorrbizal association with oaks and other trees and the systematic position of some mycorrhiza-forming fungi is unknown. Field mycologists are well aware of the common occurrence of particular species of fungi under certain trees of which this is, perhaps, one explanation. The mycorrhizal association in the orchids differs widely from that of trees. The fungi concerned are endophytes, mostly sterile *Rhizoctonia* stages of the genus *Corticium,* though the honey agaric, *Armillaria mellea,* forms mycorrhiza with the tuberous orchid *Gastrodiaelata,* in the East. In the *Rhizoctonia* type, the hyphae enter the cells of the cortex of the root, where they form dense coils which eventually break down into an amorphous mass. It has long been known that orchid seeds are difficult to germinate, and the surmounting of this difficulty by providing the seeds with cultures of a suitable strain of *Rhizoctonia* is a good example of the application of science to practice by many orchid specialists, who not infrequently have kept their methods as a closely guarded secret.

The heaths, in general, resemble the orchids, except that the endophytes, belong to the genus Phoma and in some cases e.g. in the common ling, Callunavul occur as a sparse mycelium through out the plant as far as the fruit capsules, whence the seed on germination may become naturally infected. In these two families and a few others similarly dependent on fungal infection for satisfactory germination, the seed is small and lacking in food reserves, so that it would appear that the endophytes supply some nutritional requirement. What this is not definitely known, but the fact that satisfactory germination of orchid seeds occurs without the fungus in certain sugar solutions suggests that carbohydrates are involved. In the heaths a fixation of atmospheric nitrogen has been attributed to the endophyte, but the evidence for this has been questioned.

There is much difficulty in accounting for the physiological relationships between the partners in a mycorrhizal association. Several converging lines of evidence are strongly indicative that it began as an attack by a parasite on the roots of a higher plant the prevalence of infection pegs and other reactions of the host, the occasional tendency for the fungus to develop destructive capabilities, the peculiar features of the mycorrhizal function of *Armillaria mellea,* are some of these. The eventual reduction of the invading hyphae to a condition in which they are not only made harmless but actually in large part digested by the host cells is easily reconcilable with the view that the higher of the two partners gains nutritional benefits in the process. But even this view, which is far from easy to prove, has been the subject of adverse criticism, with the result that, so far at least as concerns forest trees and orchids, the practical man was for a time more universally ready to accept it than his scientific fellows. Gradually, however, the beneficial effects of mycorrhiza have gained almost general recognition, though there is still a lack of agreement as to their nature.

The fungal partner in endotrophic mycorrhiza gains food, often using up all the starch in invaded cells, but beyond escaping the competition of ordinary soil saprophytes, seems to profit little from its colonisation of the roots. The ectotrophic forms seem to have even less chance of benefiting from the association, though it may be presumed that those of their hyphae that grow down between the host cells can absorb nutrient material from the host. There is some evidence that *Boletus elegans* gets a growth-promoting substance from the larch roots with which it forms mycorrhiza. Aneurin stimulates the growth of several mycorrhizal fungi.

An entirely different type of mycorrhizal association is found in the majority of annual and perennial plants, in which the cortex of the root is invaded, usually in limited areas behind the apical meristematic zone, by a sterile fungus characterised by special organs known as 'vesicles' and' arbuscules. The vesicles are large thick-walled diverticula or terminal swellings on an inter- or intracellular mycelium, while the arbuscules are haustoria-like, finely branched bodies formed often in particular cortical cell layers. Fat is stored in quantity in the vesicles and larger hyphae and is liberated into the host cells through the arbuscules, which collapse into formless clumps of indigestible fungal elements.

The fugi concerned are regarded as probably members of the Zygomycete family *Endogonaceae* and have been assigned to an imperfect genus, *Rhizophagus*, of this family. It has been suggested that they may supply, with the fat, some accessory nutrient or growth promoting substance to the host. The last-mentioned type of mycorrhiza is often casual in its formation and there is little evidence that it meets an obligate need of the host. But, in general, there is little doubt that mycorrhizas are beneficial and in certain cases are of great value in securing a satisfactory growth.

The lichens form the most prominent case of symbiosis in which fungi are concerned, though of lesser interest to the plant pathologist than those mentioned above. They consist of a fungus and an alga living in such perfect association that they appear to form quite definite plants. The fungus hyphae surround and grow between the algal cells, into which they sometimes send haustoria; they feed in part on the organic material which the alga builds up by photosynthesis from the carbon dioxide of the air.

The fungus in return appears to supply water, minerals and possibly peptone; it seems to have the best of matters, for it alone forms spores in most cases. The best-known symbiotic association in which bacteria take part is that of the root nodule organism of the Leguminosae, *Rhizobium or Pseudomonas radicicola.*

This is of considerable importance in agriculture, for many pulse crops will not grow well unless furnished with a suitable strain of the bacillus. The latter has the power of fixing atmospheric nitrogen and making it available to the higher plant and also indirectly to the soil when the crop decays or is ploughed under as green manure. Strains vary in their power to do this; some have been found in Great Britain that are poor nodule formers yet can outgrow and overcome the better types. It may well be that the association started in this case also as a parasitic attack by the bacillus on the root, for under certain conditions *Pseudomonas radicicola* can become parasitic. The bacillus profits by the same freedom from competition by soil saprophytes that is enjoyed by root parasites. In advanced stages of the disease dark concentric rings are visible in the flesh of the sugar beet when it is cross sectioned.

In tomato, leaflets of infected plants roll upward along the midrib and become thickened and crisp as the leaf curves downward. The leaf tissue gradually turns yellow, while the veins take on a purple tinge. The plant assumes an erect habit and remains stunted. In squash young leaves have a savoyed surface or an upward roll; internodes remain short; many blossoms fail to set; and the tips of runners turn upward. Plants infected when very young are killed rapidly without the above symptoms in warm weather.

Older vines decline gradually and die prematurely. In bean, young affected plants may show killing of the growing point and yellowing of the primary leaf, followed by death of the plant. When older plants are infected, there is a downward marginal rolling of trifoliate leaflets, followed by brittlenees, dwarfing, and puckering. The whole plant may become stunted and unproductive.

THE VIRUS

- Beet-curly-top virus.
- Sugar-beet virus 1.
- Beta virus 1 K. M. Sm.
- *Chlorogenus eutetticola* Holmes.
- *Ruga verrucosans* Carsner and Bennett.

The physical properties of the virus as determined by Severin and. Freitag and Bennett are inactivation in vitro in extracted juice, more than 7 days; by dilution, above 1-20,000; by heating, 75 to 80°C. "The virus is unusually resistant to alcohol and acetone and survives in solutions up to pH 9.1. In the phloem exudate, it withstands freezirig up to 18 months. It passes _commonly used porcelain filters. It remained active in thoroughly dried young sugar-beet plants for 8 years, leaf hoppers for 6 months, and in dried phloem exudate for to months.

DISEASE CYCLE

The virus is perpetuated between crop seasons in wild hosts and it may overwinter in the leaf hopper. In semiarid regions such as southern Idaho, abandonment of land and overgrazing often tend to encourage plants which are good overwintering hosts for the insect and which may serve as reservoirs for the virus as well. The injury to beets by transmission of the virus by the insect is dependent primarily upon the time and magnitude of spring leaf hop per movement-into cultivated fields. Factors which determine the magnitude are size of fall migration to winter feeding areas, winter survival, and spring reproduction; these in turn are influenced by the type and amount of fall, winter, and spring host plants. The date of the spring migration to beet fields is influenced by the prevailing temperature. The direction of the migration is determined by prevailing winds, which are usually from the west.

The only vector known is the beet leaf hopper; Eutettix tenellus Baker. The extract from diseased plants or infective hoppers may be used as a source of inoculum, experimentally, by allowing hoppers to feed upon it and by transferring them to the host plant. The virus has not been transmitted mechanically. Early investigators found that hoppers were not infective until 1 to 44 days after feeding on diseased plants. Sever in secured transmission in a small percentage of cases within 20 mint. Bennetit and Wallace were not successful in securing transmission in less than 4 hr., although the insect picked up the virus within 1 min. Freitag and Bennett and Wallace concluded that the virus does not multiply within the insect.

Carsner secured an attenuation of the virus by passage through *Chenopodium murale* L. Lackey showed that virulence could be restored by passing the attenuated virus through Stellaria media Cyr. Giddings isolated 10 distinct strains of the virus. He also presented evidence to show that infection was defendant upon mass action of the virus introduced.

The relation of the virus to the host tissue and the reaction of the latter to infection has been studied critically by Esau. The virus in leaves apparently is contained entirely in the phloem during the time it increases and is transported. After one or more of the primary sieve tubes have differentiated, the first histological signs appear in the form of hypertrophy and death of the pedderm and phloem cells adjacent to the sieve tubes. In cells somewhat farther removed from the sieve tubes, hyperplasia occurs; the cells involved go through some of the changes characteristic of differentiation, of sieve tubes, but normal maturation is not complete. These primary phases of hypertrophy, necrosis, and hyperplasia are followed by secondary necrosis in the abnormal sieve tubes aped their companion cells.

The parenchyma cells around the centres of secondary necrosis next undergo secondary hypertrophy and hyperplasia, resulting in preparation resembling wound periderm reactions. Hypertrophy and hyperplasia are not Confined to the phloem. The mesophyll cells adjacent to the phloem are stimulated to grow and divide so that they become closely packed together without intercellular spaces, while the chloroplasts show signs of degeneration. This internal change in the tissue is responsible for the macroscopical veinclearing noted as an early symptom. Further hypertrophy and hyperplasia in cortical tissue result information in leaf and petiole tissue, including the formation of the characteristic protuberance. Lackey demonstrated the virus, in the ineristematic tissue of the root tips.

Bennett showed that if the virus is introduced into the distal end of a beet leaf in the dark, it often fails to move out of the leaf in a period of several days, whereas when it is inoculated in a green leaf in the light, it moves out in as short a period as 4 hr. He believed that the movement of the virus has little relation to its multiplication or to its concentration but is correlated with food transport and takes place entirely in the phloem.

VARIETAL RESISTANCE

As already indicated, numerous varieties of sugar beet resistant in various degrees to curly top have been developed and released for use in various areas to which they are adapted. All such varieties are carriers of the virus and may show signs of the disease under optimum conditions.. They are genuinely tolerant, however, and the amount of phloem necrosis is greatly reduced, as shown by Bennett and Esau, while the actual amount of virus present in the tissue is less, as shown by Giddings. Lackey found that in a fairly resistant Great Northern bean the virus was promptly inactivated in the root tip, while it persisted in the susceptible Bountiful variety. Inheritance of resistance in bean has been studied by Schultz and Dean.

CONTROL

Band and crop management and controlled grazing are important measures in reducing as much as possible the magnitude of the spring migration of leaf hopper from overwintering areas to cultivated crops. Early planting and liberal fertilization of resistant varieties of sugar beet are desirable. At present certain-susceptible varieties and species must be avoided in areas where the chances of heavy infestation by viruliferous hoppers are great.

Beginning with the resistant variety of sugar beet U.S. 1, there has been developed a succession of varieties of greater agronomic value and adaptability to various areas. In southern areas, where culture comes during winter months, easy-bolting varieties must be avoided, although such a variety may be quite satisfactory for a summer crop in northern areas. The variety U.S.,22 occupied most of the Idaho acreage in 1942, and an improvement of this line was released in 1946 as U.S. 35. In the same year a new non-bolting curly-top- and downy-mildew-resistant variety U.S. 36, was released-Jar use in the San Joaquin valley and in the Imperial valley of California.

Dana apointed out that Umatella Marblehead and Yakima Marble head varieties of squash are spaturally highly olerant to curly top and are of particular value in the Pacific Northwest.

In southern Idaho the Common Red Mexican bean is resistant to curly top, while the more widely grown strains of the variety Great Northern are susceptible to curly top but resistant to common mosaic. From hybrids between these two parents a new variety of the Great Northern type resistant to both virus diseases, has been introduced as Great Northen U.I.

15. Pioneer is a curly-top-resistant snap bean introduced by the U.S. Department of Agriculture and the Oregon Experiment Station.

SPOTTED WILT

The disease known as spotted wilt was first defined as such in 1919 on tomato in South Australia, where by 1925 it had become a serious disease causing enormous losses. It has since been found oil many crop plants and wild species in continental United States, Hawaii, Great Britain, the European continent, South Africa, Brazil, Argentina, and New Zealand. Among the economic host plants besides tomato are tobacco, potato, pepper, lettuce, endive, celery, spinach, pea, and numerous ornamental species including sweet pea zinnia, nasturtium, poppy, dahlia, calla lily, orchid, chrysanthemum, petunia, aster, schizanthus, salpiglossis, cineraria, and gloxinia. The yellow spot of pineaplle destructive in, Hawaii was shown to be incited by the spotted-wilt virus. A list of hosts was compiled by Smith

The virus concerned is distinative in that it is transmitted by species of thrips, rather than by one of the more common groups of vectors of plantinfecting viruses.

SYMPTOMS

Since the disease is most widely known and commonly destructive on tomato, the symptoms on that host have been studied most extensively. In all hosts, it is well to not the symptoms may be mild under certain conditions and quite different, symptomatically and more severe in another environment. On tomato one of the first characteristic signs is bronze-coloured marking on the upper side of young leaflets, which may be accompanied by a slight downward curling of the leaves. The intensity of the bronzing varies, and it is undoubtedly influenced by the environment; it may extend to petiole, stem, pedicel, and calyx. Growth may be checked permanently or only temporarily. Upward marginal rolling and stiffening of leaflets follow, and small circular necrotic spots commonly appear on the leaves under optimum consolations. Necrosis extends to the stem near the growing tip, causing the latter to wilt and die Some mottle, chlorotic spotting, and malformation may occur on leaves. On the green fruits, yellowish spots up to 1 em. in diametre appear, often with distinct concentric zones 01 shades of yellow or bronze alternating with green and later with pink or red. The zoned fruit spots are probably the most characteristic sign of spotted wilt on tomato.

On tobacco circular necrotic spots of various sizes appear on young leaves. A pronounced mottle may follow, often confined largely to the lower half of the older leaves, while necrosis of the leaf veins adds to the irregularity of the mottle pattern. Plants are stunted, and leaves may droop gradually and die. Leaves which are mechanically show local lesions up to 4 or 5 mill. in diametre.

Symptoms vary on the numerous host plants, and often times symptoms are so variable as to be of little diagnostic value. Various types of zoned chlorotic and necrotic spots run through the symptoms syndrome.

4

Bacteria

INTRODUCTION

Since the first recognition of a bacterium, as a pathogen of plants by Burrill in 1878, many bacterial diseases of plants have been described. In the sixth edition of "Bergey's Manual of Determinative Bacteriology'" 159 species are listed as plant pathogel1s.

The bacterial plant pathogens, except for species of Streptomyces, are oreformine rdds. Their differentiation on the basis of morphology is difficult, and in early schemes of classification it was based chiefly on the number and location of flagella. The position of flagella is indicated by the following terms: monotrichous, one flagellum at one end; lopho trichous, more than one flagellum at one end; amphitrihous, one or more' flagella at each end; peritrichous, flagella scattered over the outer surface.

Even this determination was uncertain because of variability in this character in pure culture and the difficult technique involved in staining flagella. Since morphology proved to be inadequate, differentiation on the basis of physiological characters had to be employed. The most common characters used include kinds of growth on agar plates' and slants; gram stain; gelatin liquefaction; nitrate reduction; production of indole, hydrogen sulfide, and ammonia; growth in milk; acid and gas formation with various carbon sources. Here again many problems arise because of variability in cultures, difficulty in separating mixed cultures, distinction between variability and mixtures, differences between investigators in refinements of technique and interpretation of results. The understanding of variability has improved greatly in recent years. The old idea of fixed and immutable species has been abandoned. The importance of nutrition and environment in perpetuating variants is now recognized more generally. If a variant is less suited to the 'environment and substrate than the original, it may soon disappear. On the other hand a variant, and often a nonpathogenic one, may grow more rapidly than the original on a given substrate and after a few transfers completely replace the latter. In nature, vigourous pathogens in association with the host may replace less vigourous forms, and selection in the direction of pathogenicity may be going on to provide considerable stability in this character, while on a given artificial substrate pathogenicity may decline. The fact that other taxonomic character may change also makes characterization and classification the more difficult. Access to such descriptions may be had by referring to "Bergey's Manual" and to Elliott's "Manual".

Bacterial classification has gone through many changes since late in the last century, when literature on bacterial plant diseases began to appear. Therefore, the same species will be found under different Latin linomials. A comprehensive classification of bacteria was published by Migula beginning in 1897. Another classification was proposed by Lehmann and Neumann in 1896. Erwin F. Smith modified those systems somewhat for bacterial plant pathogens,' and his system is followed in Elliott's "Manual". Beginning about 1920 and continuing up to the present, a committee of the Society of American Bacteriologists has reworked bacterial classification and published "Bergey's Manual of Determinative Bacteriology." This manual has passed through six editions and it is to be expected that further revisions will in the future. It gives a list of synonyms for each species. Different investigators will undoubtedly prefer to follow different systems of classification. Some confusion, therefore, may be expected, but' the amount will be less if one who publishes on a given disease, after deciding on the binomial name he chooses to usgires also the one accepted in the system of classification 10 widely used In his country. In the united State this is the Bergey classification. A brief key indicating the relation of genera containing plant pathogens to the classification scheme of "Bergey's Manual" and the position of species treated in this book follows:

- Class Schizomycetes Nägeli.
- *Order Eubacteriales Buchanan*. Single rigid cells in chais or masses.
- *Family Pseudomonadaceae Winslow et al*: No endospores; elongate straight or spirally curved cells; usually motile by polar flagella; usually gram negative.
- *Pseudomonas*: Migula. Monotrichous or lophotrichous or nonmotile.
- *Mostly gram-negative*: 85 species listed in Bergey as plant pathogens.
- *P . solanacearum*: E. F. Sm., incitant of southern bacterial wilt.
- *P. phaseolicola* Dows: Incitant of halo blight of bean.
- *P. syringae* v.: Hall, incitant of bacterial brown spot of bean.
- *Xanthomonas Dows*: Cells usually monotrichous with yellow, water insoluble pigment; 47 species listed as plant pathogens in "Bergey's Manua1."
- *X. campestris Dows*: incitant of black rot of crucifers.
- *X. phaseoli Dows*: Incitant of common bacterial blight of bean.
- *X. malvacearum Dows:* Incitant of anular leaf spot of cotton. Family Rhizobiaceae Conn. No endospores; flagellate or nonmotile; usually gram-negative.
- *Rhizobium Frank*: Cells usually capable of fixing free nitrogen when present in root nodules of legumes; 6 species listed in "Bergey's Manual."
- *Agrobacterium Conn*: Small, short rods, typically motile; usually gramnegative; little or no gas or acid; gelatin not liquefied or sparsely so; 3 species listed as plant pathogens in "Bergey:s Manual.
- *A. tumefaciens* Conn: Incitant of crown gall of plants.

- *A. rhizogenes Conn*: Incitant of hairy root.
- *A. rubi Starr and Weiss*: Incitant of cane gall.
- *Family Corynebacteriaceae Lehmann and Neumann*: Usually nonmotile rods, frequently beaded or banded with metachromatic granules; marked diversity in form; branching tells in a few species; usually gram-positive; aerobic or anaerobic; gelatin may be liquefied, nitrates may be reduced.
- *Corynebacterium Lehmann and Neumann*: Slender! straight to slightly curved rods, with irregularly stained granules; sometimes club-shaped; snapping division forms palisade cell arrangement; usually- nonmotile;- gram-positive; 10 species listed as plant pathogens in "Bergey's ManuaL"
- *C. flalaccumfeciens Dows*: Incitant of bacterial wilt of bean
- *C. sepedonicum Skapt and Burk*: Incitant of ring rot of potato.
- *C. michiganense C. M. Jens*: Incitant of bacterial canker of tomato.
- *Family Enterobacteriaceae Rahn*: Straight rods; peritrichous flagella, or nonmotile; gram-negative.
- *Erwinia* Winslow *et al.*: Usually require organic nitrogen for growth; species vary as to carbohydrate fermention, nitrate reduction.gelatin liquefaction; all plant pathogens; 22 species listed in "Bergey's ManuaL"
- *E. amylovora Winset al*: Incitant of fire blight:
- *E. tracheiphila Holland*: Incitant of bacterial wilt of cucurbits.
- *E. carotovora Holland*: Incitant of soft rot of vegetables.
- *Order Actinomycetales*: Organisms with elongated cells tending to branch, rudimentary hyphae, in some families producing conidia.
- *Family Streptomycetaceae*: Conidia borne on sporophores.
- *Streptomyces* Waksman and Henrici. Much-branched mycelium, conidia in chains; 2 species listed as plant pathogens in "Bergey's Manual"
- *S. scabies Waksman and Henrici*: Incitant of common scab of...4) potato.

TYPES OF BACTERIAL DISEASES

The pathogenic bacteria relate themselves to their hosts in a variety of ways, and the effects on the plants differ widely. As yet none except species of Streptomyces are known to penetrate the cuticle directly, except when it has been broken or rifted by some means. The legume-nodule bacteria penetrate the non-cuticularized root hair. Many species enter through stomata, some chiefly,through hydathodes; others' enter primarily through wounds. Intercellular invasion is common. Some species continue their action primarily in the intercellular spaces. Some gain access to the vascular system and become primarily vascular parasites. The bacterial diseases may be grouped as follows on the basis of the type of major effect on the host:

SOFT ROTS

Various bacterial decays of plant organs have been described in which the major effect is as slimy softening of the tissue. Most of these are considered here as one disease, bacterial soft rot. The bacteria usually become established in wounds and advance by the secretion of an enzyme which diffuses in advance and dissolves the middle lamellae of the cells. Plasmolysis and death of the cells follow, and the IT bacteria grow upon the dead plant tissue rather than upon the living cells.

LEAF SPOTS

Many bacterial diseases start with the invasion of stomata of leaves and other organs. Invasion of and multiplication within the substomatal chamber and the intercellular spaces lead eventually to necrosis of the invaded plant tissue. Various forms of leaf and stem blight follow, primarily in the parenchyma tissue.

EXTENSIVE BLIGHTS

A more extensive and rapid necrosis is to be found as in fire blight. The progress of the bacteria is often more rapid than in leaf spots, giving a scorched appearance to affected shoots.

VASCULAR DISEASES

In some cases the leaf-spot organisms also invade the vascular system and become systemic. This is true with bean blights, which fall equally well into this and into the leaf-spot group. In other cases the invasion may be through stomata, water pores, or wounds, but the organisms promptly concentrate and multiply in the vascular system. This usually results in a stunting and often in an acute wilt common examples are in all of those mentioned the bacteria become systemic in the larger xylem vessels. In later phases of the diseases the organisms may invade xylem parenchyma and other contiguous tissues., which they are primarily intercellular. In one important disease, bacterial canker of tomato, the bacteria are contained primarily in the phloem, rather than the xylem, during the early part of disease development.

BACTERIAL GALLS

In the four groups just described the effect upon the host is primarily necrotic and/or hypoplastic. Another group consists of those in which the primary effect of the pathogen on the host is the stimulation of cell division. This condition is known as hyperplasia. Some cells become abnormally large, a condition known as hypertiophy. The abnormal condition results in over growths, which usually take the form of galls on roots, stems, or other organs. Crown gall, which affects many species of plants, is the best known and lost intensively studied disease in this group. Others are olive knot, hairy root of apple, cane gall, beet pocket rot, oleander knot, and pea fasciation. The formation of roginodules in response to nitrogen-fixing bacteria, is also a hyperplastic effect. While these organisms usually have symbiotic relation with the host, they may at times be pathogenic.

BACTERIAL SOFT ROT OF VEGETABLES

Bacterial soft rot is one of the most destructive diseases of vegetables in storage and transit. In blackleg of potato, now regarded as a phase of the same disease, the growing plant is affected, as well as the tubers before and after harvest. The disease was studied intensively in the years just preceding and following 1900 by L. R. Jones in the United States, by Harrison in Canada, by Appel in Germany, and by van Hall in Holland. The association of the causal organism with the see a corn maggot fly was worked out by Leach and first reported in 1926.

SYMPTOMS

The common name of the disease arises from the charac:" teristic soft decay of the fleshy tissues. Carrot roots, potato tubers, onion bulbs, and fleshy petioles of celery are reserve storage organs, and hence the cells are in a semidormant or "dormant condition, and the content of carbohydrate and nitrogenous materials is high. When of rot affects them, the tissue softens; it becomes watery br slimy in consistency, and as the rot progresses, the water extrudes; or in a dry atmosphere water is lost rapidly by evaporation. Some tissues, such as those of onion, cabbage, and potato, give off an offensive sulfurous odour, which may be due in part to the action of secondary bacteria.

When the seed tubers of potato are affected in the soil, the same type of decay follows. If the soil is moist for a protracted period, the shoots arising from the seed tuber are similarly affected, the cortical tissue be coming watery and the entire shoot wilting. If the soil moisture is reduced as the disease progresses, the base of the shoot develops a blackened, shriveled cortex and its growth is stunted, but it is not promptly killed. The leaflets become reddish in colour at the tips; the branches become stiffened and more upright than normal; and the entire shoot gradually becomes pale and yellowish. It is the latter phase of the disease on potato which is commonly known as blackleg.

THE CAUSAL ORGANISM

Erwinia carotovora Holland, 1923. The soft-rot group of bacteria contains organisms which are closely related in their morphological and physiological characteristics and in their serological affinities. Investigators differ as to whether they should all be regarded as members of the same species Some regard them as variants of the *ame* organism.

DISEASE CYCLE

Wounds are the most common avenues of infection. Harvest bruises, freezing injury, and insect wounds are frequent predieposing factors. The organism lives over in soil and in decayed refuse. Abundant moisture at the surface of the tissue, where wounds are present, is essential for invasion. Penetration through lenticels of freshly harvested potatoes has been observed. After infection has taken place, fairly high relative humidity is essential for progress of the disease. When decaying roots or tubers are placed in a dry atmosphere, the rotted tissue dehydrates rapidly, and further advance of the disease may be checked completely.

Several species of maggot fly are known to carry the organism. Adult flies lay eggs on decaying vegetable material, and as the eggs hatch, the larvae feed upon such debris, ingesting bacteria including the soft-rot organisms. There is good evidence that such organisms are essential to the normal development of the larvae. The intestinal tract thus becomes infested, and the pupa and the adult fly are each internally contaminated. Infested adults lay eggs in the soil just above planted Patato seed pieces, and on maturing carrot, onion, cabbage, or turnips. As such eggs are laid, they are smeared with soft-rot bacteria. Mouth parts and intestines of the young larvae become contaminated. As they feed on potato seed pieces, or fleshy organs of vegetables, they carry bacteria and infect the wounded plant tissue. The gnawing action of the larvae provides a particularly favourable and efficient means of infection. The larvae penetrate any cork barriers the plant may form. Thus the insect becomes a very important vector of the causal organism. The relation of the insect to the organism has been worked out by Leach, Johnson and Bonde.

The bacteria produce an enzyme which is capable of dissolving the middle lamella of the host tissue. This process was studied in detail by J pnes, whose work was a pioneer investigation in bacterial enzymes associated with pathogenicity. He referred to the enzyme as pectinase, but since it actually acts upon propectin in the middle lamella, it is, according to current terminology, referred to as middle-lamella propectinase. The action of this enzyme is to dissolve the middle lamella and thus to loosen the cells from one another. The by-products of bacterial growth also set up exosmosis of liquid containing sugar and salts from within the cells to the intercellular spaces, where it undoubtedly serves as a source of food for further bacterial growth.. Eventually cells in the older decayed tissues become plasmolyzed. The continuation of this process accounts for the watery condition and for the loss of consistency of the decayed tissue. Jones showed clearly that the pathogenic action was actually brought about by the diffusion of the enzyme in advance of the bacteria, while the latter grew upon liquefied lamelea and the exosmosed material.

CONTROL

In so far as soft rot of storage vegetables is concerned, the chief means of control have to do with handling of the crop at harvest. Bruising of plant parts should be avoided. Provision should be made for 1 healing of wounds and for drying of surfaces. Storage should be at the lowest temperature at which the particular product can be expected to retain its culinary quality. With the practice of washing potatoes at harvest before shipment, heavy losses are sometimes suffered in transit because of the ideal conditions for infection by soft-rot bacteria. This may be corrected by running potatoes through a warm-air current before they are I sacked. For control of potato blackleg, potato refuse piles should be destroyed to prevent them from becoming a source of inoculum.

FIRE BLIGHT OF PEAR AND APPLE

This disease is very destructive on pear and to a lesser extent on apple. Quite a number of species of the family Rosaceae including some wild and ornamental species are affected. First observed in New York before 1800, the disease has spread throughout much of the United States and Canada.

Fire blight is so severe on certain desirable varieties of pear, such as Bartlett, that commercial culture of this fruit as been abandoned in many otherwise favourable areas in the Middle West and East. The pear-growing industry has move largely to certain regions along the Pacific coast. The disease is least severe in the coastal valleys of that region, but the largest production is still in the inland Sacramento valley of.California and in the Rogue River valley of region.

The disease is of historical importance because it is the first one in which a bacterium was shown to be the inching agent. Beginning in 1878, T. J. jurrill, Professor of Botany at the University of Illinois, published a series of papers in which he reported the results of his studies showing the relation of bacteria to the disease. "some 10 years later Arthur repeated and extended the studies, using Koch's recently published pouredplate technique. It was Wait who a little later working in the U.S. Department of Agriculture, reported that honeybee, visiting the blossoms of infected branches, became ontaminated with the cteria and transmitted them to other blossoms on the same and other trees.

This was one of the first cases of experimental proof of insect transmission of organisms pathogenic on plants. Jones was the first to describe the disease on stone fruits. Fire blight is an important disease in New Zealand and Japan. Although it has: been prevalent in Italy since early in this century, and occurs in the Near East and in eastern Europe, it has not as yet become prevalent in western Europe and the British Isles.

SYMPTOMS

Fire blight appears first on blossoms, sepals; leaves, and twigs. The obsoletes turn brown, and the discolouration - extend to pedicel and often do will the spur. Leaves may be affected at the margins or on the blade, lesions ometimes being checked or sometimes enveloping the whole blade. Dieback extends down twigs and water sprouts, killing the tissue rapidly and turning it to a characteristic brown colour. The dead leaves persist, giving the appearance of having been scorched. Especially in warm humid weather, recently invaded parts may exude embroidered droplets of exudate made up of bacteria in the gelatinous matrix of plant extract and bacterial by-products. The exudate sometimes spreads to form a thin glistening layer. It dries down to hard beads.or films. Under greenhouse conditions Ivan off and Keitt observed bacterial exudate in the form of aerial strands up to 45 in diametre and several centimetres in length, consisting of bacteria and a cementing substance. While these were not found in nature, they may possibly occur sometimes. They are readily detached and tran sported by air currents.

When twig blight extends to limb, trunk, or root, localized or diffuse cankers develop, varying in size and shape. As they dry out, they may become separated from the normal tissue by rifts. They may girdle twigs and branches, and when they affect' the base of the trunk, the disease is known as collar blight. It should be recognized, however, that collar blight' may be incited by other agencies as well.

THE CAUSAL ORGANISM

Synonymy

Disease Cycle. Blighted twigs and stem cankers form the chief loci of overwintering organisms. The cankers are more consistent sources of primary inoculum in some varieties of the host than in others. The organisms are most active in the living host tissue at the outer edge of the canker, where drops of exudate form on the surface in the spring.

Waite showed that the bacteria may be carried from blossom to blossom by honeybees, and this has been regarded in some regions as the chief means of dissemination. Some investigators have shown that the concentration of the sugar and other constituents of the nectar is important in determining whether the bacteria can multiply when introduced into it.

Rain is a major factor in dissemination of the pathogen from holdover inoculum to blossoms, as shown by Gossard and Walton in Ohio and Miller in Wisconsin. Miller, in a 3-year study of dissemination on apple, found insects to be of secondary importance. He showed that if the source of inoculum was present in the upper part of the tree, the region of secondary infection brought about by the downward spread of bacteria by spattering rain was cone-shaped, a pattern which would not be explained if bees were the chief means of dissemination. If the holdover cankers were on the lower branches of the tree, the extent of secondary infection was much less. Tullis considered rain dissemination important in Michigan. In some regions water dissemination is apparently not regarded as the most important means of spread, and insect transmission is considered to be more important.

Infection takes place through the openings, which resemble stomata, in the receptacle cups of the blossom, through stomata in sepals and leaves, and through wounds. The bacteria develop first, apparently in a motile state, in the intercellular spaces of the tissue. As they increase in number, they are embedded in a matrix the nature of which is not known. Haber and Nixon believed that the bacteria migrate in zoogloeal strands, but Miller and Rosen did not agree with this interpretation, while Ivanoff and Riker showed that a similar phenomenon in crown gall was explained by the migration of liquids in intercellular spaces due to negative pressure and capillarity. Mter the exosmosis of water and solutes from the host cells, plasmolysis and relatively rapid death follow. There is evidence of some cell-wall dissolution, and bacteria are found within cells.occasionally. The rapid advance of the pathogen is, however, largely through the intercellular spaces. The progress of the disease is more rapid at 24° and 28° than at 16°C.

VARIETAL RESISTANCE

The trees of some varieties of pear are killed quite rapidly by stem and tree cankers while those of some other varieties are very resistant. Differences in resistance of varieties were recorded as early as 1817. Apples as a whole are more resistant than pears, but some apples, such as Transcendent crab and YeIlow Transparent, blight quite readily. Among the resistant varieties of pear Kieffer is the most widelyplante although the fruit is inferior in quality. It is a hybrid between Ghise Sand Pear and Bartlett, the latter being one on the choicest but one of the most susceptible commercial varieties. Under favourable conditions for the disease Kieffer often has severe blight. Richard ceters, developed in Pennsylvania, is another resistant variety.

Reimer has conducted the most extensive study of resistance in pears and wild relatives at the Southern Oregon Experiment Station. Most current varieties used in America are members of Pyrus communis L., seedlings of which are used as root stock. Most of these are very susceptible to blight, and in regions such as Oregon where trunk and root blight are severe more resistant root stocks were found in P. calleryana Decne.- A few fairly resistant varieties in communis, such as Old Home, are valuable as trunk and body stocks. Until a resistant variety desirable from tie standpoints of root hardiness, productivity, and quality is secured, a temporary solution may be had by a synthetic tree.

This is made up of a resistant root stock of P. calleryana, a body and trunk of one of the more resistant varieties such as Old Home, and a top of the desirable commercial but susceptible stock such as Bartlett. Grafts are made on several of the resistant branches, and if blight kills one down to the resistant framework, a new graft may be made in place of the one which has been killed.

In areas such as that in Oregon where humidity and rainfall are relatively low and where root and trunk blight are the most important phases of the disease, such procedures are practicable. It is doubtful, however, whether the top of such a tree would survive in middle western and eastern areas, where blossom and twig blights are the most serious phases of the disease. Thomas and Ark found resistance to be partially dominant in a cross between P. augustifoia Ait. and P. gibbsivar. yunnanensis.

Brooks grolped apple varieties in Wisconsin as follows:

- *Very susceptible*: Hyslop crab,' "Transcendent. cra Yellow Transparell McMahon, Fameuse, Wealthy.
- *Moderately susceptible*: Tolman Sweet, Black Ben Davis,

Jonathan" Salome. Moderately resistant: Mcintosh, Duchess of Oldenburg, Grimes, Orange Winter, Wolf River, Whitney crab, Stayman Winesap.
- *Resistant*: Ben Davis, Charlemoff, Delicious, Dudley, Northwestern Greening, Tetofsky.

It has been observed by many investigators that the external environment and the type of growth of the various organs have an important bearing upon the current susceptibility of any given tree. Shaw studied the nature of resistance in apple. He found that low soil nutrients, high and low soil temperatures, and high air temperature tended to make plants more resistant, since they tended to favour early cessation of growth. High aimospheric humidity, on the other hand, tended to increase susceptibility.

A critical study of intercellular humidity led him to regard this factor as a major one influencing susceptibility. Link and Wilcox and Nightingale studied the carbohydrate-nitrogen ratio of the tissue in relation to susceptibility. They found shoots high in nitrogen and low in carbohydrate very susceptible, while those in a reverse ratio of nitrogen I and carbohydrate were very resistant. However, it may well be the succulence produced with high nitrogen, rather than the carbon-nitrogen ratio, which affects the growth of the bacteria in the tissue.

Thomas and Ark found tissue high in nitrogen more susceptible in general than the nearest comparable tissue which was lower in nitrogen. Thus an inherently resistant tree may become quite susceptible when humidity is high and high-nitrogen fertilization is practiced, while an inherently susceptible variety may appear fairly resistant when air temperature is high, humidity low, and the tree is in a high carbohydrate, low-nitrogen condition. Variation between isolates of the organism from several hosts and from various parts of the United States was found by Ark.

CONTROL

A generally satisfactory control of this disease on susceptible varieties of pear and apple has not been worked out. The successful culture of very susceptible varieties of pear is limited to areas where the climate is relatively unfavourable to fire blight. Spraying with Bordeaux mixture during the blossom period gives partial control of blossom blight. Excision of blighted twigs and small cankers or treatment of cankers with a bactericide reduces primary inoculum. This procedure seems to work well on pears in California but in Wisconsin it brings about I greater succulence the next year and with it increased susceptibility: and I eventual pruning near the ground level. The regulation of soil fertilization to reduce the degree of succulence of new growth is advisable.

This means caution in the use of readily available nitrogenous fertilizer. Hildebrand and Heinecke in New York found that trees brought to good vigour in an alfalfa cover crop were less subject to blight than those on sod or under Clean culture with the application of readily available nitrogen. This was thought to be due to the mechanism whereby nitrogen was liberated by tie alfalfa and taken up by the tree. In areas where climatic conditions are favourable to fire blight the extremely susceptible varieties should be avoided. The top-working of resistant varieties of pear has been mentioned in the previous section.

ANGULAR LEAF SPOT, OR BLACK ARM, OF COTTON

Angular leaf spot was described in Alabama by Atkinson in 1891. Smith was the first to reproduce the disease and to describe the causal organism. It occurs in most if not all cotton-growing regions in the world. It is common throughout the southern states and is of major importance in the Sudan, Uganda, Trinidad, and Turkestan.

SYMPTOMS

The disease appears first on the seed leaves as they come above ground. These leaves take on a water-soaked appearance, turn black, and dry up, often leaving the young green stalk with a blackened tip. Most of such plants die. In less severely affected ones, points on the petioles and stems become water-soaked and enlarge into angulutr spots about 1 mm. in diametre. The spots often coalesce, and the leaf gradually yellows and drops. Dark lesions occur on the young stems, sometimes causing girdling and death, the phase referred to as black arm. In older stems the spots, may be callused off and leave open cankers. Yellowish bacterial exudate is common on lesions in moist weather. On bolls, gray specks grow to water-soaked circular spots, which coalesce to form irregular, large, brown, sunken areas. Bolls affected when young may drop; if older when affected, they may become distorted and the lint may be discoloured.

THE CAUSAL ORGANISM

Disease Cycle. Bacteria gain entrance to the mature seed through the basal end of the chalaza. They overwinter in this manner and as contaminants on the surface of the seed or in the lint to it. When the seed germinates; the basal cap adheres to the cotyledon and is carried above ground as the plant emerges. Volunteer seedlings are the chief source of primary inoculum when cotton. Soil and meteoric water serve as means of emanation of the bacteria to new leaves and to the stem. Infection takes place through stomata. The factors in secondary dissemination have received careful study. Faul witter showed that inseets had little importance but that wind-blown rain was the chief disseminating agent in the southeastern states. In semiarid regrons, such as Arizona irrigation water is the chief means I of secondary spread. Surface water was also found to be the chief means of secondary spread in Uganda. In some regions wind-blown infected detached leaves and wind-blown dust may serve as means of dissemination. Primary infection in the soil appears to develop more readily at the cooler end of the range at which cotton germinates and grows. Relative humidity of 85 per cent or above is essential for abundant infection. Secondary infection is favoured by high temperatures; the optimum being about 35°C.

VARIETAL RESISTANCE

Relatively little has been done in the United States in the

development of resistant varieties. All but a few of the current strains of upland cotton are quite susceptible under optimum conditions for the disease. Long-lint cottons are very susceptible, a fact which may account in part for the importance of the disease in Arizona.

The most active work on development of resistant varieties has been carried out by Knight and Clouston in the Sudan. They found in resistant Uganda two major dominant factors for resistance which are,cumulative in effect. A third factor was found in G. hirsutum var. punctatum J. B. Hutchins., Silow, and Stevens. The long-lint Sakel cotton used in the Sudan is derived from the. American Sea Island cotton. All strains of this variety are very susceptible. They crossing with Ugaflda and other resistant varieties, various resistant Sakel lines have been secured. Thus this disease is being controlled satisfactorily in some cotton areas by development of resistant varieties.

Control. J. F. Duggar of Alabama recommended in 1911 delinting the seed in concentrated sulfuric acid to control the seed-borne anthracnose organism, Glomerezla gossypii Edg. Rolfs in South Carolina applied this method successfully for angular leaf spot. Brown and Gibson devised a machine for the purpose. This still remains the most effective means of treatment, although complete of elimination of bacteria within the seed is not always accomplished. A gravity separation by passing treated seed through water and eliminating floating seed, removes a large share of infected seed. Subsequent treatment with 2 per cent Ceresan or New Improved Ceresan dust has additional value.

Since the chief source of primary inoculurll other than planted seeds is the volunteer seedlings from seeds overwintering in the ground, early killing of volunteers by cultivation is highly important. This should always occur well before the first watering in irrigated culture.. Where suitable resistant varieties are available, they constitute the most effective means of control.

BACTERIAL BLIGHTS OF BEANS

Blight of common bean was first described in the United

States by Beach in New York and by Halsted in New Jersey. The first of the causal organisms was described by E.F. Smith in 1897. The disease which it incites is now referred to as common bacterial blight, although it is not necessarily the most common one in any given situation. In 1926, Burkholder described a second bacterial blight of been, halo blight; which has much in common with the' originally described disease. In 1930, Burkholder described two other bacterial blights, which have symptoms close to those of common blight and halo blight.

These are fuscous bought and bacterial brown spot. Common-blight and halo blight are widespread in the United States and Canada I east of the Continental Divide. They occur west of the Divide but are I usually of little economic importance. They- are widespread in Europe and on other continents where beans are grown under environment favour able for the diseases. Halo blight is at present more widespread and de subtractive than common blight in the United States and in northern Europe.

Fuscous blight is incited by a variety of the causal organism of common blight. It was found first in Swizerland in 1924. It was not known in the United States until Burkholder reported it in New York in 1944. Bacterial brown spot is incited by an organism which was described first in connection with bacterial blight of lima bean and cowpea. It was shown to be a pathogen of bean by Burkholder. It is relatively rare on common bean in the United States. In 1938 its occurrence was reported in Australia.

There are several other bacterial diseases of common bean, but most of them are relatively unimportant and have little in common with the bacterial blight. One which does overlap in some degree is wilt. The pathogen, Corynebacterium flaccumfaciens Dows, is a vascular parasite, and some of the symptoms are similar to those of the bacterial blights

SYMPTOMS COMMON BLIGHT

This is first; seen as small translucent watersoaked spots leaves. The spots is from stolliatal invasion by the bacteria: do not exceed 1 II diametre as a rule. However, the interving issue may if yellow lesions of dead tissue of various size-shapes small spots appeal soon. pods of different ages. On podded varieties ororpon bean small slightly sunken spots brownish to reddish in colour, appear land often coalesce. On green pods the may 15 water-soaked at first and turn reddish with age. Water soaking in spots or streaks may appear on petioles and stems and particularly on the pulvinus at -the juncture of branch or petiole and main stem. As the disease progresses, brownish discolouration of the vascular system may become evident, and brownish surface can kersroay form on the stem in the vicinity of the first node.

This top heavy podded plant may break over at this point, giving rise to a phase of the disease known as joint rot. The seeds show various degrees of shriveling and discolouration depending upon the extent of infection. Halo blight has most of the symptoms of common blight. One striking point of difference is that around individual spots on the leaf a chlorotic zone or halo up to in diametre forms promptly. Exudate may form on pod and stem lesions with both diseases; the exudate of halo blight is cream- or silver-coloured; that of common blight is yellowish.

FUSCOUS BLIGHT

ordinarily cannot be distinguished from common blight, except by isolation and study of the causal organism. The lesions on the stem are more inclined to split the stem than in common blight. Brown-spot blight appears as water-soaked spots which remain very small and become reddish brown in colour. When the.organism becomes systemic, sunken longitudinal lesions occur on the stem. Pale brown halos, about 72 in. in diametre, sometimes appear around spots on the pod. In wilt, spotting of leaves and pods does no occur. The leaflets wilt and eventually die. A water-soaked lesion extends along the suture.of the Stem cankers similar to lose in common and halo blights may occur. Infected seed is often shriveled and discoloured. The signs the seed are not distinct although, however, to facilitate elimination of the pathoge hand picking.

Common and halo blights are important transit diseases of snap beans.

THE CAUSAL ORGANISMS

DISEASE CYCLES

The common and halo-blight organisms overwinter in diseased vines and in the seed. When seed is infected, the bacteria develop in the seed coat of the germinating seed and contaminate the surface of the expanded cotyledon, penetrating through rifts in the cuticle and progressing intercellularly until the vascular system is reached. They may then proceed as vascular pathogens throughout the large xylem tubes of the plant.

Both leaf lesions and stem cankers may arise from systemic infection. Primary local infection may also take place through stomata by inoculum arising from infected cotyldons or plant debris. Continued spread of secondary inoculum from primary lesions is carried out by wind-borne rain, dust, implements, and man and animals. The blight phase may develop rapidly under favourable conditions. Common blight is favoured by relatively high air temperatures, while halo blight thrives under relatively cool conditions.

Thus they may become epidemic in different regions and at different times of the season. Pods are infected locally or by systemic vascular invasion. Seeds are infected.by surface contamination from the pod wall or by systemic invasion through the funiculus. The bacteria migrate through the intercellular spaces of parenchyma tissue, dissolving the middle lamella slightly in advance. The affected cells collapse and are invaded.

This breakdown of the tissue leads to formation of lysogenies cavities. The xylem cells are also invaded, and the bacteria may advance systemically in the larger xylem elements. When the invaders multiply to large masses in xylem cells, they neither rupture or dissolve the cell walls and reenter the parenchyma tissues. They may emerge through stomata on the stem and, enter other stomata on the same stem after dissemination. These two organisms are therefore, both parenchyma and vascular invaders. The wilt organism is confined more strictly to the vascular, system. It enters the plant through the cotyledon from the infected seed coat or through leaf and stem wounds. It does not invade stomatal and secondary infection during the growing season is known to occur only through wounds. In all three diseases infection of the seed occurs systemically through the funiculus or through the micropyle, and the organisms are thus well protected the seed. Furthermore, infected seeds serve as the chief means of distribution and perpetuation of the pathpgens. The relation of P. syringae to lima-bean tissue is described by Beach.

HOST RANGE

The bacterial-blight organisms and the wilt organism affect various other hosts within the legume family. The range of susceptibility of some of the common legumes as reported by Burkholder is given in Table.

Table. Host Range Of Bean Bacterial-blight and -wilt Organisms

Legume	X. Phaseoli	X. Phaseoli VAR. Fuscans	P. Phaseoli cola	P. Syringae	C. Flaccum-faciens
Common bean..........	+[a]	+	+	+	+
Small lima, or Sieva, bean.	+	+	+	+	+
Scarlet runner bean......	+	+	+	0	+
Hyacinth bean..........	+	0	0	+	+
Adzuki bean............	0		0		+
Cowpea................	0	0	0	+	+
Yard-long bean..........	0	0	0	+	+
Soybean................	0	0	0	+[c]	+
Kudzu vine.............	0	0	+[b]	+	0
Broad bean............	0	0	0	+	0
Field pea............	0	0	0	0	0
White sweet clover......	0	0	0	0	0
Alfalfa...............	0		0		
White-flowered lupine...	+	0	0	0	0

[a] + = inoculation tests positive; 0 = inoculation tests negative.

[b] Reported positive by Hedges but not by Burkholder.

[c] Infection on pods only.

CONTROL

Crop rotation of about 3 years is essential and is sufficient to dispose of overwintering organisms. Sanitation with bean straw should be practiced rigidly. The use of clean seed is desirable, but such seed is often difficult to secure. Both common and halo blight occur in most seed-growing areas, although some of the irrigated regions in the Rocky Mountain states and in California have climates less favourable to the diseases than others.

Varieties vary in susceptibility to common and halo blight, but a given variety may react differently to each disease pathogen. No one are completely immune, but some are sufficiently tolorarrt to justify their use in preference to others. Lists of resistant and susceptible varieties have been published for common and halo blight by several investigators.

BLACK ROT OF CRUCIFERS

The disease was first recorded in Kentucky in 1891. The bacterial pathogin, was described from rutabaga in Iowa in 1895 by Pammel. At about this time it became destructive in numerous cabbage-growing areas in the United States and was studied intensively in Wisconsin by jusseii and in this and other states by Smith. The pathogen was one of the first of the bacteria shown to be- seed-borne; this was brought out by Harding in 1904 in studies of the disease on cabbage seed plants on Long Island. Owing undoubtedly to seed transmission of the pathogen, I the disease has been reported to occur, often destructively, in many parts of the world. It occurs on many cultivated and wild crucifers. Cabbage, cauliflower, Chinese cabbage, turnip, and rutabaga are among the cultiyated hosts commonly affected severely. Other hosts include Brussels sprouts, broccoli, kohlrabi, collard, radish, rape, charlock, black mustard, and Indian mustard.

SYMPTOMS

In most instances infection occurs through water pores at the margins of leaves, and the first signs of disease often appear at this point. Occasional lesions start at wounds caused by chewing insects. In either case the tissue turns yellow, and the chlorosis progresses towards the centre of the leaf, usually in a V-shaped area with the base of the towards the midrib. In the chlorotic tissue the veins and winglets become dark in colour, and when the lesion is held before a source of bright light, the network of black venation stands out prominently.

The vessels of the larger veins and midrib.may turn black in advance of the lesion while in the mid-ribe the tissue turns brown, dehydrates, and becomes brittle. Vascular disco oration extends to the mall) stem and proceeds upward and downward. Lesions may occur on one or both of the cotyledons, which turn yellow and drop prematurely but usually not until the organism has reached the stem through the cotyledonary node. The upward advance of vascular symptoms may proceed into upper leaves, and chlorotic lesions from systemic invasion may appear anywhere on the leaf. Leaves so invited may be stunted unlaterally. Abscission layers at the bases of effected leaves develop prematurely, and early defoliation then occurs. Cabbage and cauliflower heads are invaded and discoloured. In the rutabaga, the fleshy root is invaded from the leaves, the vascular bundles showing black discolouration and later internal breakdown. The soft-rot bacteria commonly follow into fleshy storage tissue and destruction continues as soft rot.

THE CAUSAL ORGANISM

Xanthomonas campestris Dows., 1939.

SYNONYMY

- *Bacillus campestris Pam.*, 1895.
- *Pseudomonas campestris* E. F. Sm., 1897.
- *Bacterium campestris* Chester, 1897.
- *Bacterium campestre* E. F. Sm., 1911...
- *Phytomonas campestris* Bergey et al., 1923.

The organism is a rod, with a single polar flagellu.

DISEASE CYCLE

The organism overwinters in is used plant refuse and in

seed. In either case the bacteria usually enter first through stomata at the marginal sapos of the eotyledon lobe, to which the infected seed coat is often attached and carried above ground during growth. From the substomatal chamber the bacteria progress intercellularly until they reach and invade the extremity of spiral vessels. In the Tatter they progress systematically throughout the.plant. In the leayes, stomata are not invaded, but infection occurs through water pores at leaf margins, and from they loci they advance through the intercellular spaces to the spiral vessels.

It in not uncommon for infection to occur on cabbage and cause flower in the seedbed, when cotyledons and one or two lower leaves become infected and drop off prematurely. With unfavourable conditions for further dissemination the external signs of disease may disappear for several weeks, the bacteria progressing slowly within the vascular tissue of the stem. Visible systemic lesions may appear some weeks after transplanting in the upper leaves. Bacteria which reach the surface of such lesions and are spread by spattering rain are the chief means of local dissemination.

By systemic invasion of seed stalks, the bacteria may reach the siliques internally. Seeds may thus be contaminated on the surface, but to some extent invasion of the seed coat takes place through the, vascular system in the funiculus. Mildly affected seeds are capable of germination the following season, and from their sea coats bacteria gain access to the cotyledon. After a protracted dormant period in the early part of the growing season, the disease commonly increases rapidly under favourable conditions of temperature and moisture, and epidemic development ensues. Off times an epidemic is traceable to a given lot of infected seed or to a seedbed which became infested early from inoculum in refuse or in seed. Insect transmission by larvae of the cabbage worm is known but is not important. The chief means of loca1 dissemination are wind-blown rain, cultivating implements, and transplants. Widespread distribution is by seed and transplants.

In regions where rainfall is very low during the period- when seedlings are being grown, the absence of this important disseminating factor precludes development of the disease. In the region around Puget' Sound and in many other Pacific coast areas such climate occurs. Black rot is little known and practically absent in these localities. Since most United States-grown seed of cabbage, cauliflower, turnip, and rutabaga is produced in these areas, the seed from such sources is practically always free from the black-rot organism. This is an outstanding example of producing seed free from a seed-borne pathogen by growing, the seed or in an environment which prevents infection.

CONTROL

Seedbed rotation is important. Three years absence of susceptible crucifers is necessary. Seed should be disinfected by treatment in water at. 50°C. for 30 unless it is certain that the seed is free from the black-rot and blackleg organisms:

SOUTHERN BACTERIAL WILT

Southern bacterial wilt affects a great many hosts and it has become known by many common names, including Granville wilt of tobacco and brown rot of potato. In the United States it is most prevalent in the southeastern states, particularly on tomato, tobacco, potato, eggplant, pepper, and peanut. It is a common disease of tomato in home gardens in that region. Until a resistant variety was developed, it was severe on the tobacco crops in North Carolina, where in the sandy soils of Granville County much of the good tobacco land was at one time abandoned for this crop, because of heavy infestation by the wilt patnogen.

It is a serious disease of potato, eggplant, and pepper in the coastal areas of North. Carolina, in the Hastings area of northern Florida, and elsewhere in the southeastern states. It is common ill many tropical countries. In Indonesia it is a major disease of tobacco and potato; in the Philippine Islands it is destructive on tobacco, tomato, and eggplant; in many tropical regions it affects banana. The first record of the disease appears to be on tomato in Italy in 1882. In addition to the above-mentioned countries it has been reported from Japan, India,

Ceylon, North Australia, and South Mrica. The first monographic treatment was by Smith in 1896.

SYMPTOMS

On tomato, the lower leaves may droop before wilting occurs. The vascular system becomes brown, and development of adventitious roots along the stem may be enhanced. In potato the leaflets take on a bronze colour, shrivel, and die, and leaves may wilt. The vascular system of stem and tuber turns brown. Whitish bacterial exudate oozes. From the vascular system at of cut stems and cut tubers. In damp soil the bacterial mass may exude from the eyes of the tubers, to which soil there when the crop is dug. When young tobacco seedlings are affected, the wilt develops rapidly. The older plants are stunted with less permanent wilting of the leaves, and the vascular system is discoloured. Older leaves of eggplant wilt first and young leaves later, but the disease is less severe on this host than on tomato and tobacco. Peanut foliage loses its green J. colour, wilts suddenly, and turns dull black.

THE CAUSAL ORGANISM

- *Pseudomonas solanacearum* E. F. Sm., 1914.

SYNONYMY:

- Bacillus solanacearum E. F. Sm., 1896.
- Bacillus nicotianae Uyeda, 1904.
- Bacillus sesami Malkoff, 1906.
- Pseudomonas sesami Malkoff, 1906.
- Bacillus musae Rorer, 1911.
- Bacterium solanacearum E. F. Sm., 1914.
- Bacillus musarium Zeman, 1921.
- Erwinia nicotianae Bergey et al., 1923.
- Phytomonas solanaceara Bergey et al., 1923.
- *Phytomonas ricini Archibald, 1927.*

DISEASE CYCLE

The pathogen is a soil inhabitant, persisting for long

periods of years in some spills. It invades the plant through wounds, usually below ground. Actual penetration has not been studied carefully, and it is not known whether the bacteria invade through rifts in the root cortex by secondary roots, or otherwise. The organisms migrate first in the large xylem vessels, and in succulent stems they progress into the intercellular spaces of the cortex and pith, causing lysogenize cavities. In potato tubers the organism may produce an internal delay emanating from the vascular bundles and causing a breakdown in the bud tissue of the eye.

The physiological response of the host varies with the degree of intention. Invasion of a single lateral bundle in a tomato petiole causes drooping while invasion of all bundles causes wilt. When vessels of the tomato are invaded, there is a tendency for adventitious roots to develop in the tissue just outside of the invaded bundle. The wilting is thought to be due to a gradual mechanical blocking of the xylem vessels, but it may well be that further study will reveal that toxic materials produced by the pathogen supplement the physical factors in producing symptoms.

The organism becomes established in southern soils, but the disease seldom occurs north of Virginia, although it is probably introduced into northern soils repeatedly on transplants of tomato and pepper. Potato tubers also are a means of distribution.

Relatively high soil moisture and soil temperature favour the disease. The severity of infection was increased in Indonesia by superphosphate and decreased by nitrogenous fertilizers. Smith found that nitrogenous fertilizers, particularly ureal could be used as a partial control on tobacco in some North Carolina soils, but it was necessary to use 1,000 lb per acre and to grow a crop of corn between the application and the tobacco crop in order to maintain quality, of the latter. Eddins found that increase of the hydrogen-ion concentration of the soil to pH 4 for a summer period, followed by a fall application of lime to bring the reaction to about pH 5, reduced the organisms in the soil sufficiently to be commercially practical.

CONTROL

When the pathogen becomes established in soil, satisfactory control is difficult. Eddins's sulfur-lime treatment, mentioned above, is feasible when the crop is valuable enough to warrant the expense, and it is necessarily soil types. In tobacco lands some control is accomplished by relation and fertilization. In the Granville area of North Carplina, a very successful resistant variety has been developed which has brought tobacco growing back to a profitable level in this area. Some control through resistant varieties of vegetables has been accomplished. At the Puerto Rico Experiment Station a resistant strain of tomato LJX-7, was developed by crossing Louisiana Pink with a native resistant type. Resistant eggplants have been reported in the Philippie Islands, Ceylon, and Puerto Rico. A resistant peanut is reported in Indonesia. Green Mountain and Katahdin varieties of potato were, more resistant in northern Florida that other varieties grown there.

BACTERIAL WILT OF CUCURBITS

Bacterial wilt is a common and often destructive disease on muskmelon and cucumber. Squash and pumpkin are susceptible but are not affected as severely. A number of cucurbits, including watermelon, have been in fected artificially. The disease occurs throughout the United States and Canada east of the Rocky Mountains; it is more common in the north learn states than in the south. Wilt has been reported in Europe, South Africa, and Japan.

SYMPTOMS

The first signs of wilt appear usually on individual leaves, which become flaccid at the edges in suny weather. They turn dull green in colour as the disease progresses more leaves wilt, and eventually an entire branch is affected. The wilting then becomes permanent, and the leaves and vines shrivel and die. Occasionally an exudate on fruits is visible. When, wilted stems are cross-sectioned, the viscid sticky bacterial matrix exudes from the bundles and may be drawn out in strands an inch or more. This feature is used as a means of diagnosis.

THE CAUSAL ORGANISM

- *Erwinia tracheiphila Holland, 1923.*

SYNONYMY

- Bacillus tracheiphilus E. F. Sm., 1895.
- Bacterium tracheiphilum Chester, 1897.

The organism has four to eight peritrichous flagella, and capsules are formed.

DISEASE CYCLE

The bacteria overwinter in the bodies of adult cucumber beetles, of which there are two species, the striped beetle and the spotted beetle. Primary infection is produced when beetles feed up young leaves or cotyledons. This is the only means of natural infection known. This pathogen is one of the unusual cases in which the organism is completely dependent upon the insect for its survival. After infection, the bacteria invalue and progress within then tracheal vessels. The capsular material of the organisms provides a viscid matrix in the vessels and is believed to cause wilt by means of mechanical plugging of the xylem.

Whether or not other more diffusible toxic by-products of the bacteria have a part in pathogenesis is not known, but such a possibility is to be considered. Secondary spread from primarily infected vines is carried on by beetles. The thermal death point of this organism is distinctly lower than that of other plant pathogens. Smith suggested that its sensitivity to heat might be a reason why it occurred less commonly below the latitude of Virginia.

VARIETAL RESISTANCE

There is a considerable difference between cucumber varieties in their susceptibility to wilt, but none are very highly resistant. Doolittle observed that certain European varieties tested were more susceptible than American varieties. There is come indication that varieties may be improved in resistance by breeding. As yet no highly resistant varieties of cucumber or melon have been released.

CONTROL

Since the pathogen is entirely dependent upon the beetle for its perpetuation and dissemination, control of wilt involves control of the insect. It is not enough to reduce insect damage to a commercially acceptable degree, since insects will still feed on the plants in sufficient numbers to cause an epidemic. It is essential, therefore, to apply insecticides early, to forestall the colonization of overwintering adults, and to continue applications frequently. There are good indications that a fungicide applied with the insecticide enhances the degree of control. Since cucurbits are sensitive to DDT, it should be avoided or used with caution as an insecticide on these crops.

RING ROT OF POTATO

This disease was found in Germany early in the present century, where it was described by Spieckermann It was reported in eastern Canada in 1931 and in Maine in 1932. It appeared in rapid succession in various regions of the United States and Canada. By 1940 it had been found in 37 states, and in some instances it had caused extremely heavy losses. It had not been reported in Great Britain up to 1949. The pathogen has been found in nature only one potato. It has been shown experimentally to be infectious on tomato, eggplant, and some other related species of the Solanaceae.

SYMPTOMS

In the field of the disease appears first as a diffuse and progressive pale yellow orosis of leaflets, which may flag under some conditions. Marginal necrosis appears next and progresses along with chlorosis. The entire plant shows sunting and may wilt in some cases. The chief distinctive feature, however, is chlorosis and marginal burning. Examination of the interior of the stems of plants showing advanced symptoms reveals browning of the vascular elemey squeezing, a milky viscous bacterial exudate may be forced out of the cutest. In tubers the disease sometimes may not be seen before harvest, symptoms appearing later in storage. The first sign in the tuber is a vascular discolouration, which is light yellow, becoming, brown with age. Later the bacterial ooze may be evident, and it may be intensified by squeezing the tissue. As the bacterial exudate becomes more copious, breakdown of the tissue of the vascular ring to form macroscopic lysogenize cavities follows. Dehydration then occurs, and the advanced lesions may take on a dry, powdery a pearance or they may become soft and mushy with the invasion by softrot bacteria. The vascular cavities may extend to pith and cortex and in the Jatter may result in canker lesions on the surface of the tuber. Ring rot should be compared with brown rot with which it has some overlapping symptoms.

THE CAUSAL ORGANISM

- *Corynebacterium sepedoncum* Skapt. and Burkh., 1939.

SYNONYMY

- *Bacterium sepedonicurh* Spieck. and Kotth., 1914.
- *Aplanobacter sepedonicum* E. F. Sm., 1920.
- *Phytomonas sepedonica* Magrou, 1937.
- *Corynebacterium sepedonicum* Dows., 1939.

The organism is nonmotile.

DISEASE CYCLE

The rapid spread of this disease in North America was associated with dissemination of seed tubers from special growing areas in northern states and Canada. The organism lives from season to season chiefly in tubers. It will subsist also in the dried slime on crates, bins, harvesting and grading machinery, and sacks. There is no evidence that it persists in the soil. However, volunteer plants from surviving tubers in the soil may be a source of primary inoculum. The initial invasion is through wounds. There is no evidence of invasion through stomata, lenticels, or water pores. Since the bacterium is resistant to drying in the slime mass, contaminated machinery and cutting knives are the chief source of primary inoculum, which enters through cut seed pieces and through harvest and handling bruises. There is some evidence of spread,in the field by irrigation water and from hill to hill through the root system. The disease becomes established in the large vessels of the -vascular system of stem and tuber and progresses in this channel systemically. Eventually the bacterial mass-progresses into xylem parenchyma and contiguous tissue, where lysogenize cavities are formed. The bacteria may remain for a long time in tubers without producing visible symptoms.

It is possible for a tuber to carry the germs, internally, and produce symptomless plants and symptomless tubers although all three stages are infected. This is especially true when growth of the host is under continuously cool environment Larsoll and Walker showed that no top symptoms appeared in plants from infected tubers grown an, temperature of 16°C., while stunting vas severe at 24°C. Sherf secured maximum infection of stolons and tubers in plants growing at temperature of 18°. Experimental transmission through tomato, seed has been demonstrated.

CONTROL

Since the bacterial-ring-rot organism is distributed on seed tubers, rigid of certification standards followed its appearance in the United States. In most states no visibly infected plants of tubers are permitted in seed stock to be certified. Since final field or warehouse diagnosis is not always possible, supplementary laboratory procedures are sometimes necessary.

When a grower once has ring–rot–infected stock on his property, thorough disinfection of all warehouses, bins sorting equipment, and cultivation machinery becomes necessary because to the extreme infectiousness of the pathology arid its dissemination by equipment. Used bags should be discarded or thoroughly disinfected. For bins, crates, and bags copper sulfate is best. For machinery formaldehyde or BK is recommended. Following such sanitary precautions, only foundation or certified seed stock should be used for planting. All other potato stocks must be kept off the premises.

Since the organism is transmitted from tuber to tuber by the cutting knife, one infected tuber in a lot may lead to a general infection of seed stock by this means. Use of rotary knives which run continuously through 1-500 mercuric chloride has been urged by many investigators. Frequent renewal of chemical disinfectants is necessary to assure favourable results. Continuously boiling water is as effective as chemicals if it is kept at boiling temperature. Another device is a stationary double-edged knife over which 1-500 mercuric chloride flows continually. Picker planters facilitate greater spread from seed piece to seed piece than do assist-feed planters.

Some progress has been made in the development of resistant strains. One new resistant variety, Teton, has been introduced. It is well to point out that this is not immune and is successful because it produces satisfactorily in spite of infection. It still remains a possible carrier of the pathogen and thus is a potential source of contamination where susceptible varieties are being grown.

BACTERIAL CANKER OF TOMATO

The disease was first described in 1909 by Erwin F. Smith on tomatoes growing in a greenhouse at Grand. Rapids, Michigan. He first called it Grand Rapids disease but later named it bacterial canker. It was found in a few northern states in the following 18 years but did not assume major importance until 1926 when it caused considerable damage on outdoor tomatoes in western New York. In the years immediately following, it became a major disease in field-grown tomatoes throughout the canning belt from New Jersey to California and in several southern states.

It has caused losses in many other countries including Australia, New Zealand, South Africa, Kenya, Morocco, Germany, Russia, England, and Canada. The disease has been found in nature only upon tomato, except for one instance reported from southern California in which stem cankers were found on perennial nightshade in a tomato field. The isolate from nightshade was infectious on tomato and on nightshade, but an isolate from tomato was not infectious on nightshade. Ark produced infection by artificial inoculation on tree tomato on red-currant tomato on *Solanum nigrum var. gu.ineense L., and on Nicotiana glutinosa L.*

SYMPTOMS

Temporary, and later permanent, wilting of leaflets appears at the extremities of plants of any size. The next signs are light streaks beginning at the juncture of leaf petiole and stem, extending down the internode and up the petiole. As these streaks become darker, the epidermal tissue may rift, and a longitudinal canker-like opening may appear in stem, petiole, or midrib.

There may be a stunting of the plant, or occasionally young plants may appear to recover as they grow older. Leaves may be affected unilaterally, the leaflets on one side of the rachis being stunted or wilted while those on the opposite side appear healthy. When the stem of a diseased plant is cut lengthwise, a creamy white, yellow, or brown line follows the phloem. Cavities in the older diseased tissue extend into the pith or into the cortex and to the -outer surface. The discoloured vascular system can be traced to the fruit, where it becomes brown and, ramifying to the outer surface, may cause brown blister-like exterior lesions. The discolouration may also extend to the placentae and the seeds.

The symptoms described above appear as the result of systemic infection by the pathogen. Secondary spread from canker lesions is common. Invasion takes place through stomata on leaves, cotyledons, and stems, small light-coloured spots forming which are not often conspicuous in nature. Secondary infection appears on green fruits as water-soaked spots which soon become surrounded with a white halo.

The spots remain less than 3 mm. in diametre exclusive of the halo. The presence of a halo has been emphasized by Bryan as a means of distinction between the canker spot and that of bacterial spot incited by another organism, *Xanthomonas vesicatorium* Dows, which is reported not to have a halo. This is not entirely correct, since the latter disease commonly does have a halo surrounding the fruit lesion in the early stages, and at this time the two diseases are indistinguishable on green fruits. In bacterial spot the halos disappear as the lesion becomes older.

THE CAUSAL ORGANISM

- *Corynebacterium michiganense* Jensen, 1934.

SYNONYMY

- *Bacterium michiganense* E. F. Sm., 1910.
- *Pseudomonas michiganense* Stevens, 1913.
- *Aplanobacter michiganense* E. F. Sm., 1914.
- *Phytomonas michiganense* Bergey et al., 1923.

The organism is nonmotile.

DISEASE CYCLE

The organism lives over from season to season within the seed coat. It may also be carried on the surface of seeds. Blood discovered that when, the fruit pulp is allowed to ferment at room temperature for 72 hr. or more, the bacteria on the surface and within the seed coat are, as a, rule, completely destroyed. Fermentation of pulp in the commercial extraction of tomato seed has long been a standard procedure, and this fact accounts in a large measure for the minor importance of the disease until 1926.

About the latter date, cyclonic machines for extraction of tomato juice were introduced in canning factories. Since it was possible to save both juice and seed by this process, the need for fermentation in seed extraction was removed. However, the omission of fermentation allowed perpetuation of the canker organism, and seed became the means of rapid and widespread dissemination. As a result, the disease was elevated from a minor to a major role in the United States, and it began to appear in many other countries, possibly for the same reason.

The seed coat is commonly attached to the cotyledon and carried above ground after germination. Invasion of the cotyledon through stomata is commonly the locus of primary infection. Signs of the disease may then disappear, as in black rot of crucifers, while the organism grows slowly in the vascular system. Reappearance of symptoms may come after seedlings are transplanted. In any case they usually show as wilting of leaflets and later as open cankers. Local dissemination is from open cankers and is carried out chiefly by spattering rain. Seed becomes infected internally by systemic invasion of the fruits. Widespread distribution of the pathogen is by seed and by transplants.

The organism becomes established early in the phloem. It is the only known vascular bacterial organism which migrates through the plant primarily in the phloem. Eventually it invades the cortex and causes lysogenize cavities, which may break open at the surface to cause cankers.

The disease is influenced by various soil and air factors. It may develop at air and soil temperature as low as 16° and as high as 36°. The optimum soil and air temperetures for disease development are about 28°C. This is also about the best temperature for growth of the tomato. However, if plants are, grown at this temperature or slightly below it before they are infected, they resist the progress of the disease more than plants predisposed at higher or at lower temperatures. Walker and Kendrick showed that after infection the rate of disease development at the optimum temperature was enhanced considerably by increasing the concentration of the nutrient solution up to a certain level.

CONTROL

Bryan did not succeed in ridding infected seed of the organism by chemical or by hot-water treatment. The best method is fermentation of the seed and pulp at room temperature for 72. hr. before extraction. Blood showed that in the process of extraction the hydrogen-ion concentration of the juice rose considerably; and he surmised that high acidity was responsible for eradication of the bacteria. He devised a soak in acetic acid, which is about as effective as fermentation if it is applied to freshly extracted seed. In addition to fermentation much reliance has been placed in recent years on inspection and certification of seed fields.

There is little evidence of overwintering of the organism

in crop refuse. As a precautionary measure for this and other tomato pathogens, sterilization or rotation of seedbeds and rotation of the main field are advised.

CROWN GALL, HAIRY ROOT, AND CANE GALL

In the bacterial diseases discussed in the previous pages the major effect on the host plant is necrotic and/or hypoplastic. Crown gall is representative of a group of -diseases in which the major effects on the host are hyperplasia and hypertrophy. This disease affects a considerable number of host plants in nature, and a much wider range of plants has been shown to be susceptible by experimental inoculation. Crown gall is important on apple, peach, apricot, almond, grape, rose, and cane fruits.

In 1897, Cavara, in Italy, first isolated a bacterial organism from galls on stems of grape and reproduced the disease; by inoculation. Numerous other investigators worked with the disease in Europe and America between 1892 and 1904 without successfully isolating an inciting agency. Erwin F. Smith and C. O. Townsend took up a study of the disease in 1904. The pathogenicity of the causal organism was demonstrated by 1906, and the organism was described in 1907.

In 1930, Riker distinguished hairy root from crown gall and described the causal organism of the former. It has a more limited host range than the crown-gall organism. It was reported to be pathogenic on apple, rose, honeysuckle, sugar beet, bean, and Paris daisy, but very slightly so or not at all on tomato and tobacco.

Cane gall of brambles was distinguished from crown gall by Banfield in 1934. Hildebrand described the pathogen as a distinct species in 1940. It is confined in its pathogenicity to members of the genus Rubus.

Crown gall has been extensively studied because of the opportunity to examine basic causes of pathological growth. The extensive considerations bearing on the cancer problem are too voluminous for discussion here. Reviews by Riker et should be consulted.

Symptoms. Crown gall consists in over-growths varying

from the size of a pea or smaller to 50 lb. or more. Galls occur commonly on the sub terranean roots of trees or shrubs. They may also appear 9n, crowns, stems, or leaves of woody and herbaceous plants. The over, growth may be relatively soft and thus is readily disintegrated by secondary organisms; it may consist of a corky exterior and a hard woody interior.

Hairy root, or woolly knot, commonly occurs on grafted nursery apple trees. The distinctive symptoms are excessive production of small roots, sometimes from hard swellings which often occur at the graft Unions of piece-root-grafted apple trees.

Cane gall appears as small protuberances or elongate ridges on the fruiting canes of Rubus spp., somewhat before the middle of the growing sead. These increase in size and may completely cover sections of the surface, especially in the lower portion of the cane. The gall tissue is whitish at first, turning brown with age., As the season advances, severely affected canes split open the leaves wither, and the berries dry up.

Wound over growths, non-patasitic in nature are confused commonly with crown gall since they may occur along with the latter. Such over growths often occur on apple trees at the margin of root and scion, particularly when an imperfect union has been made.

Mixtures of crown gall, hairy root, and' wound overgrowth are commonly found in the nursery.

THE CAUSAL ORGANISMS

Crown-gall organism: *Agrobacterium tumefaciens* Copn, 1942.

SYNONYMY

- *Bacterium tumefaciens* E. F. Sm. and Towns., 1907.
- *Pseudomonas tumefaciens* Stevens, 1913.
- *Bacillus tumefaciens* Holland, 1920.
- *Phytomonas tumefaciens* Bergey et al., 1923.

The organism has one to four polar flagella:

HairY-foot organism: Agrobacterium' thizogenes Conn, 1942.

SYNONYMY

- *Bacterium rhizogenes* Riker et al., 1930.
- *Phytomonas rhizogenes* Riker et al., 1930.
- *Pseudomonas rhizogenes* Riker et al., 1930.

A. *rhizogenes differs from A.. tumefaciens* in the following ways: (a) it is unable to utilize such simple nitrbgero 1S compounds as potassium nitrate; (b) it absorbs cango red and bromothymol blue slightly and aniline blue not at all; (c) it is unable to grow on sodium selenite agar; and (d) it is able to form acid from various sugars.

- *Cane-gall organism*: *Agrobactrium rubi.*
- *Phytomonas rubi Hildebrand,* 1940.
- *Bacterium'rubi Hildebrand,* 1940.
- *PseudQmonas rubi Hildebrand,* 1940.

A. *rubi* differs from A. *tumefac!ens* in that it does hot utilize nitrates and grows much more slowly on ordinary media. It differs from both *A. tumefaciens* and *A. rhizogene*s in that it does not utilize asparagine as a sole source of carbon and nitrogen.

DISEASE CYCLE

The organisms are successful soil inhabitants. They are distributed widely on living planting stock. Infection takes place entirely, so far as is now known, through one or another type of wound. Subterranean chewing insects are important in facilitating: crown-gall infection. Hildebrand found that, in general, the deeper the wound, the better the chances of infection and the larger the gall.

Smith in an early publication described the bacteria as intracellular. However, Riker and Robinson and Walkden found the organisms to be intercellular like other gall-forming bacteria. Bacteria may be carried a short distance from the wound by liquid which fills the intercellular spaces, but usually no farther than inert material would be transported by negative pressure and capillarity. Rapid cell division takes place around the point where bacteria are present in the intercellular spaces. Hyperplasia and hypertrophy of cells follow. An old gall consists of a mass of such tissue and often elements which resemble scalariform cells of the vascular system.

In 1912, Smith reported secondary tumors, which were connected by strands of tumer tissue, and interpreted his findings as comparable to metastases in animal cancer. This view has been doubted by others. Riker and Robinson and Walkden showed that if the crown-gall bacteria were inserted in wounds near the growing tip of a plant like Paris daisy or sunflower, they might follow the released intercellular liquid through a number of nodes. As the growing tissue elongated rapidly, they would later be found to be distributed in several disconnected loci, at each of which a gall developed. Thus Riker found in one case with sunflower that secondary galls appeared 13 nodes and 49 cm. distant from the point of inoculation. Most-workers since Smith agree that tumor strands often fail to connect primary and secondary galls. Braun claimed the development of sterile secondary tumors and tumor strands in sunflower and secured sterile transmissible tumor tissue from secondary tumors, but it is possible that the bacterium originally present may have died.

In studies on crown gall of tomato Riker found the organism to grow on culture medium from about 10 to 34°C. with an optimum at about 22°. The host grew best at from 22 up to 30° soil temperature. When stems were inoculated below the soil line with crown-gall bacteria, gall development was best at 22°, meager at 28 to 30° and nil above 30°. Gall development in this case was greatest at the optimum for the pathogen and some what below the optimum for the host.

Although most bacterial over growths in apple trees in the nursery occur at the graft union, Riker found that new infections occurred in wounds during the second, third, and fourth seasons after the graft was made. Infection by the hairy-root organism subsequent to the healing of the union seemed to be correlated with the occurrence of warm weather, with the activity of soil insects, and with rapid growth of the tree. The incubation period for both crown gall and hairy root was relatively long in the spring and fall and relatively short in the summer.

Extensive physiological and biochemical studies have been made with the organism and upon the fundamental factors underlying atypical growth in the pathological tissues. These are reviewed by Riker et al.

E. F. Smith and Quirk described a species of Begonia immune to the crown-gall organism. C. O. Smith found a wide range of resistant and susceptible species and varieties within the genus Prunus. Some of these have value as root stocks. Wormald has described four physiologic races of the crown-gall organism on the basis of host reaction. The gall inducing bacteria were compared by Riker et al.

CONTROL

The chief approach to control of crown gall, hairy root, and cane gall is through the production of clean nursery stock. Obviously little can be done after the orchard has been set. Inspection and certification is required in all states before removal and this to a large extent keeps these diseases in abeyance.

For control in the nursery, certain definite procedures are necessary. Proper disposal of infected stock is paramount. Clean areas where rootchewing insects are not prevalent should be selected for the nursery. Wellfitted grafts wrapped with antiseptic tape, together with sanitary practices in grafting and storage, must be had to ensure a minimum of infection and to reduce the amount of confusion with around over growths. Antiseptic solutions should never be applied to antiseptic tape wrappers, since they unduly delay or prevent suitable decay of the cloth and girdling of the plant may follow.

Ark has reported effective eradication of galls on almond with elgetol painted on the stem galls and an inch-wide zone of surrounding bark. Brown and Boyle have shown that soft galls of *Bryophyllum* disintegrated when penicillin was applied to them. They suggested this as a possible means of control.

COMMON SCAB OF POTATO

There are two scab diseases of potato which are distinguished in popular terminology as common scab and powdery scab. Common scab has been known for more than 100 years in both Europe and America. The causal relation of the inciting organism was first shown by Thaxter at the Connecticut Agricultural Experimental Station in 1890. The greatest loss from scab is brought about by the blemish to the appearance of the tuber and the consequent reduction in commercial grade. When lesions are deeply pitted, the as the in peeling is increased. Predisposition of the tuber to chewing insects adds to the loss. The disease is world-wide in occurrence. In highly acid soils it is relatively scarce, while in soil with moderate hydrogen-ion concentration, and particulady in muck soils, it is most prevalent and most important.

SYMPTOMS

Shallow and deep scab are two phases commonly referred to in popular and scientific literature. Shallow scab consists in a superficial roughened area, sometimes raised above, and often slightly below, the plane of the healthy skin. It consists of corky tissue which arises from abnormal proliferation of 'the cells of the periderm of the tuber, resulting from invasion by the pathogen. The lesions vary widely in size and shape and usually are slightly darker than, but not greatly different in colour from, the healthy skin. This symptom may be such that it, causes a diffuse russet, or it may grade into the deep phase. Deep scab, or pitted scab, consists of lesions which are 1 to 3 or more mm. deep. They may be darker, than shallow lesions. The tissue around the interior of the pits is corky, as in shallow scab. Opinions differ as to the nature and origin of deep-scab lesions. One view is that certain physiological races of the organism incite this type of lesions while other races incite shallow scab, Another view is that deep scab is the result of combined action of the scab organism and of chewing insects attracted to the incipient lesions. Relatively inconspicuous lesions may be found on stolons and on subterranean stems. Lesions on roots of other plants in special cultures are described by Hooker.

THE CAUSAL ORGANISM

- *Streptomyces, scabies* Waksman and Henrici, 1948.

SYNONYMY

- *Oospora scabies* Thaxt.., 1890.
- *Actinomyces scabies* Güssow, 1914.

There are two prevalent points of view concerning the systematic position of the, group to which the scab organism belongs.

Drechsler came to the above opinion after a thorough study of the morphological characteristics of this group. The opposite view is taken by others who believe that the actinomycetes are closely related to the higher bacteria, in which rudimentary mycelium occurs. In the sixth edition of "Bergey's Manual", the forms which produce conidia' in chains on aerial hyphae are placed in the genus Streptomyce.s. The potato-scab organism and the incitant of soil rot of sweet potato, *S. ipomoea* Waksman and Henrici, are the only species in the order which are recognized as plant pathogens.

Aerial mycelium in pure culture consists of prostrate branched,threads. Sporogenous hyphae are spiral in form. Spores are produced by the formation of septa at intervals along the hypha, which contract to form narrow isthmuses between the cells. When the spores in the chain mature, final cleavage takes place. The spores are roughly cylindrical and hyaline. They germinate by means of one or two germ tubes.

DISEASE CYCLE

The pathogen may subsist indefinitely in favourable soil. It is distributed chiefly on infected seed tubers and in wild- and water-borne soil. It may- survive passage through the digestive tract of animals and become distributed with manure.

The young tuber is most susceptible to infection. The pathogen enters chiefly through lenticels but also through stomata, wounds, or directly through the cuticle when the latter is very scanty. It progresses intercllularly at first, apparently digesting the middle lamella. The invasion results in the stimulation" of cambium cells to become -hypertrophied and the walls to become suberized. As the fungus progresses, activity continues until a lesion is formed. The extent and type of reaction varies with the host variety, with the, environmental conditions, and possibly with the strain of the parasite. The larvae of soil insects are common invaders of scab lesions, and they are responsible, in part at least, for the formation of deep lesions.

Various soil factors influence the severity of the- disease. One of these which has received much study is soil reaction. Numerous experiments have shown that the extent of infection is reduced quite rapidly as the pH decreases below about 5.2. There is some evidence that above about pH - 8.0 infection is also inhibited. Within the range pH 5.2 to 8 the severity of the disease is little influenced, although it tends to increase somewhat up to about pH 7.0. There are numerous records of some infection at pH 5.0 or less. This has been explained by Some as due to the presence of a strain of the pathogen highly tolerant to acid. However, it is rather generally accepted that when potato soils are kept at pH 5.0 to 5.2, the disease is usually under commercially satisfactory control.

Shapavalov found that the spores germinated most rapidly at 35 to, 40°0., while, the cardinal temperatures for growth were maximum, 40.5°; optimum, 25 to 30°; minimum, 5°. Jones in a study of the relation of soil temperature to disease development found the range to be between 11 ° and 30.5° with an optimum at about 20 to 22°. Sanford noted in 1923 that increase in soil moisture tended to reduce scab, and this was confirmed by Martin. Dippenaar studied the relation of temperature, moisture, and soil reaction on scab development. He found the optimum temperature for infection to be slightly below 20° an hat for lesion enlargement to be slightly above 20°. He also found that, cab decreased with increase in soil moisture. Goss emphasized the importance of these soil factors and of aeration in their effect on the soil flora for a period before infection. In some experiments the largest amount of scab occurred in soil held at high moisture content for some months before infection. Lack of soil aeration for a period before infection depressed the disease more than poor aeration provided during the post-infection period. Starr et al. in Wyoming, in a study of the effect of different irrigation schedules on incidence of scab, found most disease in the schedule which gave the high~ total water supply and the least variation of soil moisture.

The reduction of scab by certain rotation schedules has been noted by some workers. Millard and Taylor in England emphasized the value of green manure as a scab depressant. They and Sanford were of the opinion that the pathogen was inhibited by the activity of competitive organisms. Goss and Manasiev compared the effects of rotations on' scab under irrigation in western Nebraska. They found that 4-or 6-year rotations with alfalfa were most satisfactory.

It appears at present that soil reaction and soil moisture are the most influential factors on seasonal and regional incidence of scab. But it is equally apparent that the interaction of these with other less clearly defined factors makes too rigid generalization unwise. By and large, scab remains an important disease because clear-cut, workable, reliable soil-management proyeduresare,not yet defined.

Waksman pointed out that some strains of the pathogen grew in culture media at pH values as low as 4.8. Millard and Burr noted variability in pathogenicity of strains of the scab pathogen. They set up several new species and believed that some caused shallow and others deep scab. Bruyn, in Holland, was inclined to the same view. Afanasiev, in Nebraska, was of the opinion that the production of shallow, russet, and deep scab was a function of the degree of virulence rather than of type of pathogenicity.

Leach described two physiologic races on the basis of reaction to a seedling and to the resistant variety, Jubel. The seedling was very susceptible to race 1 and highly resistant to race 2, while Jubel was only moderately susceptible to race 1 and highly resistant to race 2. Schaal showed that most isolates are unstable in pure culture, variants being produced frequently which differ in physiology, morphology, and pathogenicity. This was confirmed by the work of Thomas. Taylor and Decker reported a close correlation, in 143 isolates studied, between pathogenicity and formation of a dark brown ring of surface growth on milk. It is obvious that the scab pathogen is an exceedingly variable organism; this fact.may account in part for the complexity of its reaction to environmental conditions.

VARIETAL SUSCEPTIBILITY

For a long time it has been known that varieties differ in their susceptibility to scab. In some strains in which tubers have russeted skin fewer lesions appear than in smooth strains of the same variety, e.g., Russet Rural, Russet Burbank and Russet Sebago.

There is reason to believe, however, that the apparent resistance is not due to a smaller number of infections but to the fact that the reaction of the cells near the surface is more limited and the lesions are thus less conspicuous. Some varieties developed in Germany and other European countries, e.g., Jubel, Hindenburg, Ostragis, and Arnica, are very highly resistant. They are of no value in American potato culture as such but have been 'used widely as parents by potato breeders. Scab-resistant introductions derived from those parents include Menominee, Ontario, Seneca, and Cayuga.

Longree found a correlation of resistance with degree of suberization of the lenticel meristem and with a tendency of the maturing tuber to slough the lesion. Darling found the lenticels of susceptible seedlings larger and made up of rounder and more loosely arranged cells than those of resistant seedlings, while the periderm of the latter suberized earlier and the suberization extended farther into the lenticel. Clark et al found some varieties, e.g., Green Mountain; to breed true for susceptibility while Hindenburg and Ostragis were apparently homozygous for resistance.

The susceptible variety Katahdin carries at least one gene for resistance in a heterozygous condition. Russeting was found to be linked genetically with resistance. Stevenson tested 22 varieties in scab gardens in Maine, Michigan, Minnesota, and Colorado during two successive seasons and found the resistance and susceptibility reactions to be fairly constant, indicating that variability in the pathogen would not seriously hamper the development of, resistant varieties.

CONTROL

Treatment of potato seed tubers for control of scab was initiated by Bolley in North Dakota in 1891. Many experiments on tuber treatment have been conducted in various parts of the United States and elsewhere. Formaldehyde and mercuric chloride have been used extensively. The original formulae usually recommended were 1-1,000 mercuric chloride soak for 1½ to 2 hr., or formaldehyde 1-240 soak for 2 hr. In order to shorten the period of treatment, the hot formaldehyde treatment and the acid-mercury dip were devised. The danger of tuber or sprout injury is greater in the shorter methods. Organic-mercury dips have been devised also, Semesan Bel being the most common. While tuber treatment is effective in reducing or eliminating introduction of inoculum, the organism has become so widely prevalent in many commercial potato soils that the benefit of tuber treatment is often of doubtful value.

Where soil is naturally rather highly acid, scab is commonly held in abeyance by use of acid-tolerant crops in rotation and avoidance of the use of lime beyond that amount' which would raise the pH above 5.2. The use of sulfur on soils with higher pH has been shown to reduce scab, but except in very special cases it has not been adopted as a commercial procedure.

The effect of rotation and green manuring has been discussed in a previous section. In a few situations procedures have been worked out which are reliable, but in the main this type of remedial measure finds little application.

Resistant varieties are the ultimate solution for control of this disease. The tolerant varieties are useful in this regard, but they are not successful under conditions extremely favourable for scab. The scab-resistant introductions listed above are all late-maturing varieties, and the extent to which they may fit into potato production has not been determined.

Scab remains one of the most important and least satisfactorily controlled of potato diseases.

5

Virus Diseases

HISTORY OF VIRUS DISEASES

There is no reason to suppose that virus diseases are of more recent origin than those caused by fungi or bacteria. The increased attention that they have attracted of late is probably rather the result of improved diagnosis and of the ever increasing traffic in plants which has led to the wide dissemination of so many plant diseases from one country to another, than of the appearance of new disease producing agents. Many virus diseases cause a progressive deterioration, with chlorosis and other obscure symptoms, and so have been more readily overlooked than the rusts or smuts or blights that injure crops.

However, one disease now known to be caused by a virus, the' breaking' of produces such beautiful effects in the flowers as to have been promptly noticed when it reached western Europe, possibly from Turkey, in the sixteenth century. It was described by the celebrated botanist de l'Ecluse in 1576 and can be seen depicted in the paintings of Dutch and Flemish artists in the late sixteenth and seventeenth centuries. Methods of inducing 'breaking' in tulips were advocated, some of them highly fanciful; in 1670 there was a suggestion that the variegation might be the result of a disease, and in the eighteenth century Van Osten claimed that a whole-coloured 'breeder' could be broken by grafting a piece of a broken bulb on to it. This clue was not followed up, and there is little indication for another 250 years that broken tulips were thought to be diseased.

A similar state of mind must have existed among those horticulturists who knew more than a century ago that variegation in ornamental plants could sometimes be communicated by grafting. Even a variegation in the apple was transmitted by Vibert in France by shield-budding in 1863. In the elaborate scientific studies of the variegation which is characteristic of the *Abutilon* variety called *Thompsoni* by horticulturists transmission by grafting to normal green plants was fully established between 19°4 and 1908. This type of 'infectious chlorosis', as it was termed, was found to affect various other Malvaceae and unrelated plants but as it was only transmissible by grafting and seemed to do little harm, it was not generally classed as a disease.

The first destructive outbreak of what are now known to have been virus diseases of a crop, of which records exist, appears to have been the 'leaf roll' of the potato, which was extremely prevalent in Great Britain for many years from about 1770 onwards. A similar condition was well known in Germany about 1785. 'Leaf roll, originally known as, curl', was doubtless due to several viruses, often more than one in the same plant. The symptoms of generalised deterioration - pallor, deformity of the top, leaf and stem weakness, reduction in cropping vigour, and the like - led to the widespread belief that it was the result of 'running out' of the plant owing to long-continued vegetative propagation; degeneration and senile decay were held to be sufficient to account for the trouble.

Even as early as 1802, however, a suggestion is found that greenflies were implicated; now amply confirmed, this suggestion received little support at the time~ and the view that degeneration in the potato and other plants could best be prevented by outbreeding and the production of seedlings by fertilization of the flowers held the field for more than a century. During this period it was noticed that degeneration progressed much more rapidly in some localities than in others, and the practice became established of importing potato seed tubers from Scotland for growing in England, with very beneficial results. It was not easy to account for this with the knowledge then available, but northern hardiness and the benefits of 'change of air' were vaguely thought to play some part.

During the nineteenth century several 'other important crop diseases now known to be due to virus agencies were recognised. Among these was the' sereh J disease of sugarcane seen in Java in 1882; this rapidly became a menace to the important and increasing sugar industry of the island and was the cause of the development of biological, chemical, and cultural research into cane problems which led to effective control of sereh and secured for the Dutch East Indies the leading place in sugarcane research.

'In 1890 mosaic of sugarcane was first mentioned under the name' yellow stripe disease' in Java; unlike sereh, it subsequently spread to all the cane-growing countries of the world, and the extensive deterioration it caused makes its discovery a landmark in the history of plant virus diseases. Tomato mosaic was mentioned in England in 1887 and described five years later in the United States.

Meanwhile another disease, this time of tobacco, began to attract attention in northern Europe. It was described by Mayer of Holland in 1886' and termed, mosaic" disease, a name based on the marbled effect of light and dark green patches on the leaves which found popular favour and was afterwards applied to many diseases causing leaf mottling.} Mayer showed that tobacco mosaic could be mechanically transmitted from a diseased to a healthy plant, reproducing the disease.

Further studies of the same disease by I wanowsky in 1892 demonstrated that infection of a healthy plant by the juice from a diseased one still occurred after the juice had been filtered through a porcelain filter that prevented the passage of bacteria, and also that the infective matter multiplied abundantly in its new host plant, resisted drying and alcohol, and thus differed from bacteria and the like; therefore he thought the infective matter must be something in solution or perhaps in minute bodies in the cell plasma. F or the first time the existence of invisible disease-producing agencies smaller than anything yet known was postulated; their study was made possible and a basis for virus research established. A few years later the famous Dutch bacteriologist Beijerinck, knowing nothing apparently of I wanowsky's work, made a detailed study of Mayer,'s disease and, reaching the same conclusions as Iwanowsky, propounded in 1898 his theory of the *contagium vivum fluidum,* an infectious living fluid or non-particulate agent of disease! It was in the same year that Loeffler and Frosch, working on footand-mouth disease, first demonstrated that animal diseases occurred of the type now known to be due to viruses.

Soon after Mayer's work Erwin F. Smith carried out studies of a destructive disease known as peach yellows in the United States and in 1888 published the results of experiments proving that the disease could be transmitted by grafting or budding but not by any other of the methods that he tried. Beijerinck thought that peach yellows was another example of the *contagium vivum fluidum,* and it was later recognised as being allied to tobacco mosaic and similar diseases.

For a number of years after Beijerinck's work there was little progress in the study of virus diseases of plants though there were isolated investigations of interest. 'Thus, in 1901 the Japanese Takami succeeded in transmitting the 'dwarf' or 'stunt' disease of rice by the leaf hopper *Nephotettix apicalis,* the first insect transmission of a disease now known to be due to a virus on record. In 1902 Kamerling claimed to have transmitted sugarcane mosaic in Java by injecting sap from a diseased into a healthy cane.

Between 1906 and 1915 a group of workers in the United States associated beet' leaf curl' with infestation by leaf hoppers, then showed that these insects caused the disease, and finally that they could only do so after feeding on a diseased plant and that the infectious agent was not merely carried mechanically by the insect. In 1913 Quanjer, who had already started his classical researches on the virus diseases of potatoes in Holland, reported that a necrosis and lignification of the phloem was a constant concomitant of potato leaf curl and was an infallible symptom of the disease, which he therefore termed 'phloem necrosis' of the potato. In 1914 Orton described and named potato mosaic which he had seen in Germany in 1911, and introduced.the name' leaf roll' instead of 'curl' into American and English usage, and also the term' streak' for yet another type of these potato diseases. From 1914 to 1918 Allard greatly extended knowledge of tobacco mosaic. In 1916 Quanjer proved that potato leaf roll was an infectious communicable disease.

It was still possible, however, for one of us to write in 1918 in discussing tobacco mosaic "It is perhaps, the best known example of a group of diseases of obscure origin, in which not only has no parasite been detected, but there is strong reason for believing that none of any type known to us can exist. Some hold that it is a physiological disorder, due to an innate alteration in the processes or functions of the plant, not brought about by the presence of an independently living parasitic organism; others, that it must be parasitic but due to a form smaller than any yet known, of even to a 'living contagious fluid' - a suggestion which goes still further beyond the limit of any life known to us. It seems more reasonable to suppose, as other investigators have, that the virus is one of an enzymic nature, capable of producing extensive effects in minimum doses... Similar diseases are know in tomatoes; chillies, and other plants.... They form a group which is perhaps the most obscure in the whole field of vegetable pathology."

Soon afterwards the successful insect transmission of spinach blight and potato 'mosaic' in the United States, of potato leaf roll in Holland, and of sugarcane mosaic in the United States, together with the, discovery in the United States that cucumber and bean mosaics were transmissible through the seed, marked the initiation of a period of intense activity in plant virus research.

Investigation of potato virus diseases was actively prosecuted in Holland, the United States, Canada, Great Britain, Ireland, France, and many other countries. Records of diseases of the 'leaf curl' and' mosaic' types became frequent, so that in 1921 it was stated that mosaic was known in 30 genera of 10 families of plants and mosiac-like diseases in 8 genera of 5 families. "The view that these diseases were caused by ultramicroscopic organisms was gaining in credence. Studies of their behaviour in the insect host, especially in *Eutettix tenellus* the vector of curly top of sugar beet in the United. States, were throwing light upon the relation between insect and virus. By this time, too, the view that 'degeneration' was a concomitant of long-continued propagation from the same stock had been rudely shaken.

The opinion was definitely expressed in England in 192 I that potato "degeneration is but a symptom of a disease, and this disease is probably mosaic". In France in the same year it was stated that degeneration is only the consequence of disease or poor adaptation of a plant to its environment and that, in the potato at least, propagation by asexual means cannot be held to lead inevitably to degeneration of a variety; while in the United States, again in 1921, the running out of a large series of potato varieties observed in Minnesota was believed to be due to the introduction of mosaic with some of the tubers brought to the Experiment Station in 1883.

THE NATURE OF THE VIRUS

Throughout this period and for some time afterwards attempts were made to identify as the causal organisms of virus diseases of plants certain elements seen in the diseased tissues or obtained in cultures from them. In some cases bodies reminiscent of the 'elementary corpuscles' described in certain animal and human virus diseases were thought to be the infective organism, and the name *Strongyloplasma ivanovskii* was given to bodies seen in tobacco mosaic.

Amoeboid bodies found in maize and *Hippeastrum mosaic* and, later, in those of sugarcane, tobacco, and *Brassica* were thought possibly to be protozoa or other living organisms causative of the diseases, as some held the Negri bodies in animals affected with rabies or the Guarnieri corpuscles associated with smallpox to be Flagellates and Trypanosome-like organisms were reported in the phloem of various plants suffering from virus diseases; but were later shown to be normal constituents of the cells. The new genus *Phytamoeba* was established for supposed amoeba-like organisms seen in Fiji disease of sugarcane, while *Plasmodiophora tabaci* was reported in mosaic tobacco plants, and minute motile organisms were stated to occur in the cells and chloroplasts of mosaic tomato and other plants, but were later seen in healthy plants also and were presumably due to contamination of the material.

In 1922 Twort reported his constant failure to establish cultures of ultramicroscopic organisms occuring saprophytically. He supposed that more primitive forms of life than the bacteria or protozoa' must exist and that the filter-passing lysins associated with many bacteria which he had found in 1915 may be in this class; he inclined to the view that a precellular form of life, situated in a sense between the simple enzymes and the bacteria, may constitute the ultra-microscopic viruses. Another possibility which he discussed in the following year was that the viruses might be unimolecular stages of the host which for some reason preserved their independence when the fertilised cell first divided.

Akin to this view was one expressed in the United States in 1923 that the agent might be a product of the cell such as a particle of chromatin or possibly some structure such as a gene which had escaped from the shackles of coordination in the 'cell. 'Again it was thought in 1926 to be conceivable that a virus might be an intermediate stage in the evolutionary path from the non-living to the living and might have acquired some of the properties of living matter, e.g. that of reproduction, but not others'. Though successful cultivation in vitro of the causal agent of tobacco mosaic was reported in 1924, subsequent work failed to substantiate this claim. The inclusion bodies found in the cells of many virus-infected plants were early considered to be nothing more than degeneration products of the cell, and later work indicates that they are the result of reaction or-the cell to the presence of the virus and are an effect, not the cause, of the "disease.

Up to 1935 most biologists regarded the viruses as living organisms: Though few believed that any plant virus had been seen under the microscope or cultivated artificially in cell-free media; there were still some, however, who maintained that an autocatalytic as opposed to an organic nature best explained their properties. In that year Stanley announced in the United States that he had isolated from tobacco mosaic juice a crystalline protein having all the properties of the tobacco mosaic virus; he considered the virus to be an autocatalytic protein, presumably requiring the presence of living cells for its multiplication.

The following year he reported the preparation of a similar substance from the juice of tomatoes infected by tobacco mosaic, while Bawden and co-workers who had been engaged at Cambridge in the purification of certain viruses, obtained liquid crystalline proteins from three types of tobacco mosaic which were infective at high dilutions—or even less—and could be transmitted from plant to plant in series, showing rapid multiplication in the infected plants. 'These proteins agreed with Stanley's preparations in their high molecular weight, and X-ray studies indicated that they consisted of closely packed elongated rods.

'Further studies showed them to be nucleoproteins, separating as a jelly on high-speed centrifuging, and the view was expressed, in agreement with the American workers, that these heavy proteins, which were not found in healthy plants, might well be the actual viruses; this view was strengthened by studies of potato virus 'X' reported in 1938, which showed that the virus, when purified, appeared to be a solid mass of nucleoprotein showing under X-ray examination a perfect internal regularity in the particles, of the type found in some large protein molecules.

The rods found in the tobacco mosaic and some other viruses are not true crystals, having 2-dimensional but not 3-dimensional regularity, and are better termed paracrystals purification reduces their filter-passing facility, a result apparently due to the aggregation of smaller particles present in the sap into longer rods; as shown by electron microscope 280 ± 8.6 mμ is a most common length. Later studies of the virus of the bushy stunt' disease of tomatoes yielded nucleoproteins in a form more closely approaching that of true crystals, with particles approxiniately isodiametric in shape and not tending to aggregate on purification; the diametre of these particles is about 27.4 mμ and their molecular weight has been variously calculated as from 8,800,000 to 12,800,00. In the United States, Stanley isolated and studied a heavy nucleoprotein from tobacco affected by the ring spot virus and assigned to it a molecular weight of 3,400,000 and a spherical shape, with a particle diametre of 19 mil; the tobacco necrosis virus has a particle size between 13 and 20 mil, and the lucerne mosaic virus 16.5 mfL in diametre, these being about the smallest plant viruses so far reported.

This ring-spot virus is thrown down in isotropic pellets without crystalline structure on high-speed centrifuging and can reproduce the disease.' Apparent recovery takes place sometimes, the new leaves coming out without visible symptoms; it was found that in these unmarked leaves only about one part virus in 500,000 of fresh green leaf occurred, as against one part in 80,000 in leaves bearing many ring-spot lesions. Recovery in this case is apparently merely due to a reduction in concentration of the virus in the host The concentration of the virus, however, naturally varies according to the time allowed for its multiplication in the plant; ordinary tobacco mosaic protein has been estimated to increase over a million times in a four-day period, and reaches its maximum concentration five weeks after inoculation. I have concentration reached may be one part of heavy tobacco mosaic protein in 500 parts of plant tissue.

Specific nucleoproteins have now been isolated from a number of plant virus diseases.' Some are infective in dilutions of one-thousand-millionth of a gram per c.c.; in some, the molecular weight of the order of 20,000,000, and still heavier animal virus proteins, have been reported. In the other direction a heavy protein, having a molecular weight of 500,000, has been isolated from the bacteriophages, and these are now regarded as causing virus diseases of bacteria; they range in size from 10 to 75 mil, are specific to particular bacteria, and are capable of multiplication on them. 'Each virus has its own particular protein irrespective of the host plant from which it is obtained—tobacco mosaic nucleoprotein is the same in anyone of several quite unrelated hosts, while potato X virus inoculated into tobacco produces a protein differing from that of ordinary tobacco mosaic and agreeing with the X virus obtained from potatoes. Their properties of stability, resistance to heat and other agents, and so on, agree with those that had already been established for the virus in the plant. Even strain differences in a virus can be recognised in the purified product. In the highly purified products there is very little else than nucleic acid and protein; no salts or carbohydrates or bound water, such as occur in all organisms hitherto known, have been detected.

Workers with animal viruses have had some difficulty in accepting all the implications drawn from the botanical work. It so happens that animal viruses exist that are large enough to form a descending series in size from some of the smaller bacteria, without the considerable gap that is apparent between the bacteria and the plant viruses; indeed some animal viruses are larger than some of the organisms, like that causing bovine pleuropneumonia, that have been artificially cultivated in cell-free media. Furthermore, saprophytic viruses have been discovered by animal virus workers; one from London sewage is cultivable on ordinary media. At the moment it is hard to say more than that some viruses seem to have the characters of minute organisms, others to be devoid of 'living' properties as the term is ordinarily understood. Much turns on the definition of 'living matter', but those who are most closely in touch with the recent work on plant viruses seem to be least inclined to regard them as organisms.

DIAGNOSIS OF VIRUS DISEASES

No exact criterion exists for the diagnosis of a plant virus disease. None of the symptoms is exclusive, and there is no specific property by which the existence of a virus disease can be recognised/The negative characters in the pathogen, of invisibility by ordinary methods of microscopic examination using visible light and uncultivability apart from the living tissues, are those most made use of in diagnosis, and if, in addition, the juice is found to be infect after filtration through porcelain or other bacteria-proof filters and the disease can be transmitted by sucking insects or by grafting, it is fairly certain to be due to a virus.

Not all virus diseases of plants, however, have been found to be caused by filter passers, and where the only known method of transmission is by grafting, as in 'paracrinkle' of the potato,, big bud' of the tomato, and' spike disease' of sandalwood, neither juice infection nor insect transmission can be used to confirm the diagnosis, though there is little doubt that most diseases of this class will be found in time, as peach yellows was found after many years, to be insect borne. Transmission from plant to plant in series implies multiplication of the infective agent in the host and is a valuable help in diagnosis when no visible cultivable pathogen, such as a fungus or bacterium, can be detected.

The commonest symptom of virus diseases is probably the mottling in different shades of green or green and yellow on which the term 'mosaic' is based. This symptom, however, is also found in various deficiency diseases and occasionally a genetical chlorophyll deficiency is expressed in a similar manner.' Sometimes, both in virus diseases and these other disorders, the chlorosis or yellowing is generalised over the whole leaf surface. In many mosaic diseases the pallid or chlorotic areas produced on some of their host plants are arranged in concentric circles or in lines.

A notable feature of some virus diseases of the potato, tomato, tobacco, cauliflower, and other plants is the development of dark-green bands bounding the veins; this condition is known as 'vein banding' and may be evanescent. In 'bunchy top' of bananas the most reliable diagnostic symptom is the green streaking found alongside the veins in the leaves and sheaths, due to the phloem being replaced by a morbid tissue surrounded by ground cells rich in chlorophyll. A leaf symptom of value sometimes in diagnosis is a translucent appearance of the veins termed' vein clearing this may be a very early symptom as in lily mosaic and the tobacco and tomato diseases caused by potato virus, Y "in which it is often followed a few days later by vein banding. Like the latter it is evanescent.

More characteristic than mottling are various types of curling or crinkling of the leaves, though these again may be simulated by insect attacks or, more rarely, by those of parasitic fungi. The latter, however, are usually more localised than the effects of virus diseases, which tend to be generalised throughout the plant and are sometimes fully systemic.

Necrosis results from the attacks of a good many plant viruses it may be restricted to spots on the leaf may be of the 'ring spot' type, or arranged in angular lines that have been described as resembling hieroglyphics, or may occupy small or large continuous areas; or the necrosis may occur in streaks on the leaves or stems, the leaf veins sometimes being the first to be affected. In its most severe forms necrosis affects the whole tip of a shoot, or the petioles of the leaves, or an entire branch system, causing the collapse of the parts above the area affected. Necrotic areas may also develop in storage organs, as in the net necrosis of potato tubers which results from the destruction of the phloem in plants affected by leaf roll. The tobacco necrosis virus can kill the cells of all organs except the root but never causes mottling or chlorosis; ordinary tobacco mosaic, on the other hand, only rarely, as at high temperatures, kills tobacco cells.

Distortions and malformations of organs and abnormal growths such as galls and leafy enations are evidences of the profound changes in the internal order of the plant brought about by some virus diseases. A remarkable example is the, fern leaf' type of foliage found sometimes in tomato plants infected by tobacco mosaic or by cucumber virus 'I'. In extreme cases the leaf lamina may be so reduced that the organ resembles a tendril. V:Elongation of the stems is checked in many virus disease's to several of which the descriptive popular names of 'dwarf', 'stunt', and the like have been applied. 'Bunchy top' and 'rosette' are other descriptive names of virus diseases characterized by shortening of the internodes and crowding of the leaves at the tip of the shoot. Less often, increase in size results from virus infection, as in 'giant hill' and witches 'broom of potatoes, the overgrown tomato plants produced by tomato nation virus, and the abnormally large leaves found In elm mosaic.

A characteristic feature of potato witches' broom is the growth of many thin wiry shoots from the leafaxils and base of the plant; these may sometimes be threadlike and leafless. In 'spike' disease of sandalwood the new leaves formed after infection is established are narrow, pointed; stiff, erect, and crowded together by shortening of the internodes; they become progressively smaller and narrower, so that eventually the new shoots resemble fine spikes bearing four rows of bristles. Wiry, finely branched shoots with very narrow leaves arise from the trunk and branches of peach trees suffering from yellows.

Galls are caused by a few virus diseases, the best known of which are the' big bud' of tomatoes and Fiji disease of sugarcane; the latter has as its chief symptom the production of elongated swellings along the veins of the under side of the leaves. In the' stripe' disease of narcissus and daffodils small galls or corrugations occur on the leaf surface, while small warts are found on the veins of the lower leaf surface in sugar-beet' curly top'. Thickening and increased prominence of the veins are found io several virus diseases, especially of the 'curl' or 'crinkle' type. In leaf curl of tobacco and cotton the vein thickenings are green and are often associated with the formation of cup-shaped or laminar leafy outgrowths known as 'enations '.

Enations also result from infection of peas or broad beans by enation pea mosaic, and of tomato and some species of *Nicotiana* by tobacco mosaic in the British Isles, North America, and elsewhere. A characteristic spindle-shaped or cylindrical swelling of the shoots, due mainly to hyperplasy of the xylem, is a late symptom of the swollen shoot disease of cacao, epidemic in a part of the Gold Coast.

The floral abnormalities caused by virus diseases include the variegation of the petals so beautifully shown by many' broken' tulips and similar striping of the petals in mosaic narcissus, iris, stock, wallflower, sweet pea chrysanthemum, peach, and so forth. Phyllody occurs in aster yellows, spike disease of sandalwood, and big bud of tomatoes and beans. In 'false blossom' of cranberry all the floral organs are abnormal and the flower may be replaced by whorls of leaves or a small branch.

The well-known, teratological prolongation of the axis of the affected flowers termed median floral proliferation is sometimes seen in 'aster yellows' and' big bud' of the tomato. Complete sterility sometimes results from failure of the stamens to mature, while up to 50 per cent of the pollen has been found to be abortive in tobacco; or it may be due to abnormalities of the gynoecium as in tomato big bud. Plants affected by the sugar-beet curly top virus may have twisted peduncles, dwarfed and desiccated flowers sometimes lacking the corolla, and malformed or aborted fruits and seeds. Flowering is largely suppressed by cucumber mosaic and aster yellows in- some hosts; or young flowers and fruit may be lost by abscission or by drying up.

In groundnut, 'rosette' flowering may occur but no nuts develop after infection. Similarly whole fields of susceptible varieties of cotton have been seen in the Sudan where early total infection with leaf curl had not seriously impaired vegetative vigour, but it was difficult to find any mature bolls. Early blossoming is a feature of some potato and tobacco virus diseases and early fruit maturity of peach yellows.

The roots of plants are little affected by the presence of virus diseases. In sugar-beet curly top there is an increase in the number of thread roots and in some hosts, such as tomato, the rootlets decay backwards from the tip. 'Several viruses have been successfully cultivated in the excised root tips of tomato for long periods without the slightest external symptom appearing.

The susceptibility of some crops, such as potato and tobacco, to infection by several different viruses was early recognised. The discovery in 193 I that simultaneous- infection with more than one virus caused some of the well-known potato virus diseases, and the analysis and synthesis of these diseases, marked a notable advance in virus pathological research. At an earlier date it had been shown experimentally that inoculation of tomatoes with a combination of potato and tobacco mosaics caused a very severe 'streak disease', while evidence was obtained that two tobacco diseases in Kentucky were each caused by a mixture of two viruses. It is now fully established that some of the most severe diseases of the potato and other plants are composite mosaic diseases; potato crinkle is due to a combination of potato virus A with X, potato net necrosis to A plus the virus of potato leaf roll, while tomato mixed streak is caused by tobacco mosaic virus plus potato virus X. Another serious complication in the diagnosis of these diseases is the great differences in the symptoms they induce in different varieties of the host plant.

This has nullified attempts to base a classification on symptoms. The most complete studies of varietal reaction to infection by particular viruses have been made on potatoes, and the literature on these diseases is loaded with detailed descriptions of their effects on the numerous varieties commonly cultivated. Apart altogether from the symptomless' carriers' of a virus and the phenomena of resistance to, and tolerance of, virus infection, varieties that are apparently fully susceptible may react in different ways to an individual virus. Tests in Holland of the reaction of a large number of varieties to infection by the' Y, virus served to distinguish five types of reaction: the potato varieties in group 'A' suffered from mosaic with drooping of the leaf tips and margins; in group' B' there was, in addition, a certain amount of rugosity with a few fine veinal necroses and yellowing or bronzing of the lower leaves; group 'C' added to these symptoms a streak disease causing leaf drooping and desiccation owing to necrosis extending back from the leaf veins to the petiole; in group 'D' the necrosis extended to the surface of the stem, while varieties in group 'E' showed little or no mosaic but there was severe dark necrosis of the veins with some necrotic spotting of the leaf, the necrosis spreading extensively in the stem and being accompanied by leaf drop.

A somewhat similar grouping has been made of the "reaction of several varieties in the British Isles to the same virus. Tuber necrosis is a prominent symptom in some potatoes affected by 'aucuba mosaic', while in other varieties there is none, though the leaf symptom of brilliant yellow spotting may be equally apparent in both groups.

As might be expected, still greater differences than those found within the varieties of a single species of host plant occur when the virus is tested on susceptible plants belonging to different species, genera, or families. These differences are greater than those observed when pathogenic fungi or bacteria are similarly tested, so much so as to raise the question whether they are not due to an entirely different order of phenomena. The common 'tobacco mosaic', for instance, which is fully systemic in the cultivated Nicotiana tabacum, produces only local circumscribed lesions on inoculated leaves of *N. glutinosa* and does not spread to the rest of the plant; similarly circumscribed lesions are produced by it in French beans, but it is fully systemic in spinach, tomato, and several other hosts. 'Tobacco necrosis virus' occurs naturally In the roots of a wide range of hosts In glass-houses at Cambridge without causing any visible symptoms; natural infection of the leaves is unusual but has been seen in tobacco, while artificial inoculation causes only local infection, except in French bean, in which it may eventually become systemic.

Tomato spotted wilt', which has probably the widest host range of any virus disease in England and affects monocotyledons as well as dicotyledons, produces every kind of symptom from local lesions without systemic infection to complete systemic necrosis which kills the plant: vein clearing, mosaic mottling, bronzing, bring spots, leaf curl, stunting, rosetting, necrotic spotting, vein and petiole necrosis, necrotic striping of the stem, distortion of leaves, flowers, and fruit, suppression of flowering, are all amongst the symptoms found in different I hosts, though only a selection of them is usually found on anyone host.

The potato viruses seem seldom to pass outside the *Solanaceae*, but some of them show widely different symptoms on particular genera. 'Aster yellows' in the United States has a host range of over 160 species in many families, amongst which many distinctive changes in habit, overgrowth dwarfing, deformation and pigmentation of leaves, floral abnormalities, necrosis, and decay have been observed; chlorosis or yellowing without mottling is, however, a general symptom. Artificial infection tests in the United States, reported in 1940, showed that cucumber mosaic virus could infect no less than 191 species in 40 families. The diseases caused by this virus or by distinct strains of it are so diverse in their symptoms as often to defy diagnosis without extensive testing or serological study; almost the whole range of symptom expression of virus infection may be found in some plant or other infected by it.

CARRIERS

The existence of 'carriers' of virus diseases, that is, of plants which contain a virus without any visible signs of infection, was early established. Thus, it was stated in 1923 that the juice from certain tomato plants that showed no symptoms two months after they had been inoculated with mosaic was capable in every case of causing visible infection of healthy susceptible plants. In the same year the existence of a carrier of hop mosaic was demonstrated. Similar results with potato roll and mosaic were reported in the following year.

Later it was found that all the main cultivated varieties of potatoes in the United States were so universally infected with certain viruses that it was impossible without special precautions to obtain a virus-free tuber; for a time the name 'healthy potato virus' was applied to the infective agent in these cases, but eventually it was recognised that. American potato varieties were all carriers of one or more latent viruses, the commonest of which were strains of the 'X' virus; the vein banding viruses 'A' and 'Y' and a top necrosis virus' B 'are also sometimes concerned.

In the British Isles, Up-to-Date potatoes always carry 'X', usually accompanied by, B '; Golden Wonder and Irish Chieftain always carry 'A' and King Edward, 'E' since 'E' is only transmitted by grafting, cannot spread naturally, and has been found in King Edward in all parts of the world, the stock must have been infected at its very origin. The 'Y' potato virus is reported to be carried masked by turnips and other brassicas as well as by red clover, garden peas, and bindweed; tomatoes show the early symptoms of infection but become symptomless carriers as growth proceeds. In experiments at Cambridge leaf roll was carried by turnips, Brussels sprouts, stocks, and campanula.

Carriers of the' leaf roll' and' Y, viruses, usually the two most destructive in the south of England, are rare in the potato varieties grown in Great Britain except in old-standing 'chronic' infections, but tolerant varieties are known in Germany, and the Swedish variety Imperial is a perfect carrier of leaf roll. Carriers of yellow edge and crinkle of strawberries are fairly common, especially in varieties in which the parent species Fragaria chiloensis is predominant, while in raspberries the well-known Lloyd George variety is sometimes a symptomless carrier of mosaic.

In all these cases the presence of the virus in the carrier is revealed by infecting by grafting or otherwise a susceptible or 'indicator' plant in which visible symptoms are produced. If potatoes carrying 'A' or 'X' virus are grafted with one or other of several potato varieties, e.g. Up-to-pate, a lethal top necrosis results and the plant is killed. The wild strawberry *Fragaria vesca* similiarly serves as an indicator plant for yellow edge and crinkle of strawberries.

The economic significance of carriers is twofold. In the first place, a carrier variety may suffer little from the presence of the disease, and under certain conditions the use of carrier varieties may mini mise the losses in a crop. This is not of universal application, however, for cases have been reported in which yields have been substantially depressed by the unsuspected presence of a symptomless virus.

Secondly, the carrier, while giving complete satisfaction in certain districts, may on transfer to a new locality become exposed to infection by another virus and the combination may be far more destructive than either alone; thus, when a strain of Xproducing no symptoms on the potato variety President becomes combined with 'A' which alone causes only a slight mottle on this variety, the result is a severe disease of the crinkle type.

The rapid degeneration from crinkle of Up-to-Date potatoes which are extensively exported from the British Isles into Southern Rhodesia is probably to be traced to the carried virus which is present in all the imported stocks meeting with another crinkle constituent found locally.

TRANSMISSION OF VIRUS DISEASES

In the transmission of these diseases it appears to be necessary for the virus to come into contact with living cell plasma. Unbroken cell membranes ordinarily oppose its passage, though there have been successful infections when a virus suspension of tobacco mosaic or of potato virus 'X'is sprayed on the surface of the leaf; these two viruses are readily transmitted by contact, as for instance when leaves of healthy and diseased plants are blown together by the wind, and it is difficult to exclude the possibility of infection through broken hairs or other surface micro-lesions, though the experiments indicate that stomatal infection is involved.

Natural infection by direct transmission of the infective agent through the air, as in most fungal and some bacterial diseases has not been found in the viruses except in the very aberrant tobacco necrosis virus which has been recovered from the air and water in infected glass-houses and reaches the roots of the host plants through the soil when every other source of contamination but air or water has been excluded. 'This virus enters the roots through broken root hairs and other natural injuries and will not penetrate roots growing in contaminated water cultures unless these are wounded.' transmission through the soil also occurs in the wheat rosette virus in the United States, while the virus of ordinary tobacco mosaic survives in the debris in or on the soil from a diseased tobacco crop, and infection from this source is the most important cause of the disease in certain areas on land on which an infected crop was grown the previous year; it has even been stated that this virus can persist in soil from one season to the next after being leached out of the infected tissues.

In tobacco and tomato mosaic, natural infection frequently results from the handling of healthy after diseased plants, as in the operations of topping and priming of the crop, the hands of the workers becoming contaminated by the infective juice from' the already diseased plants.

Tobacco mosaic virus also survives for years in manufactured tobacco, and it has been proved that not only the fingers of smokers of infective tobacco but also the sputum from those who chew the tobacco, can, and often do, infect the crop. This virus is intensely infective, successful inoculation having been accomplished by cutting with sterile scissors the hairs on infected leaves without injury to the cells below and using the scissors thus contaminated to cut hairs on a healthy leaf; even at a dilution of one part in a million it is still infective. It is also extraordinarily resistant to drying and putrefaction, some virulence having been found still persisting in 1937 in tobacco cured in 1882, while bottled juice that had undergone putrefaction remained highly infective after fifteen months. Tests in 1940 of crushed mosaic tobacco leaves kept since 1925 in water in flasks gave only a very small number of successful infections, but similar material to which the preservatives benzene or xylene had been added caused 3° to 4° per cent of the number of infections caused by a fresh extract. The inactivation of tobacco mosaic virus caused by the growth of aerobic micro-organsims is much greater than that caused by anaerobes.

Evidently it is difficult to avoid contamination of the crop with a virus such as tobacco mosaic, and both it and tomato streak have been found to be transmitted to healthy tomato plants on pruning knives, as has also sugarcane mosaic on the knives used for cutting the setts.

Fortunately there are not many viruses with such properties, and many lose their power to infect quite soon after removal from the living host; tomato spotted wilt survives in expressed sap only for about four hours and cannot stand drying. Natural spread by mechanical means can be of little practical importance in most virus diseases, for it is usually necessary to inoculate by needle-pricks or through surfaces abraded by carborundum and the like to ensure successful infection.

The transmission of plant viruses by insects is of far greater importance than, the methods already mentioned and accounts for the natural spread of the great majority of these diseases; in a recent list 70 plant viruses are stated to be so transmitted. Most of the insects concerned are leaf suckers of the order *Hemiptera*, a few are leaf abraders of the order *Thysanoptera*.

There is a small number of cases of transmission by leaf-eating insects such as grasshoppers and Diabrotica beetles and one or two by mites. Amongst the *Hemiptera*, aphides, leaf hoppers, and white-flies are the chief vectors, but there are a few plant bugs. In closely allied species, one may be a successful vector while another fails even though it may be able to pick up the virus. It is of interest to note that some viruses are rendered inactive if ingested or passed into the blood of non-vector insects and that even in the vector itself there is evidence that the virus may come into contact during feeding with some inactivating substance, for it has been found that the insect can retain the virus longer when fasting than when allowed to feed.

Aphides are the most important vectors, both from the number of viruses they transmit and their prevalence. Some aphides can transmit several viruses, the peach aphis *Myzus persicae* more than twenty; conversely, some virus diseases can be transmitted by more than one species of aphis, yellow dwarf of onions by more than fifty.

Leaf hoppers are responsible for the spread of several major diseases in overseas countries - curly top of sugar-beet, celery, aster, and peach yellows, rice dwarf, maize streak, Fiji disease of sugarcane. White-flies transmit cotton and tobacco leaf curl and cassava mosaic. The chief disease in the British Isles transmitted by insects other than greenflies is the spotted wilt of tomatoes and many other plants, of which the local vector is the Thysanopterous Thrips tabaci. The plant bug Piesma quadrata is the vector of sugar-beet crinkle in Europe. The leaf-sucking insects feed by a long and delicate stylet apparatus or beak which not only serves to suck up the leaf juices but also to inject saliva through a second channel along the passage forced through the tissues; this often reaches the phloem. It is generally assumed that the infective agent sucked up with the juice into the insect is eventually passed into the new host with the saliva, though difficulties in accepting this view have been pointed out.

A socalled incubation or latent period in the insect between the time of ingestion of the virus and its successful transmission to a new host is found in a group of virus diseases, especially those transmitted by leaf hoppers and thrips, but including a few that are aphis-borne. The latent period may be nine days or more in *Eutettix* tenellus, Macrosteles divisus and *Macropsis trimaculata,* the vectors, respectively, of sugar-beet curly top, aster yellows, and peach yellows. All the evidence available indicates that, at least in the group showing this latent period, the virus must pass from the intestine into the blood. Whether it multiplies in the insect is still in doubt.

On the one hand there are experiments indicating that the aster yellows virus can multiply at least a hundredfold in the body of its vector, and it has also been thought that sugar-beet curly top virus increases in *Eutettix,* though, if there is an increase, it is not sufficient to maintain for long the amount originally picked up from the plant van the other hand, the delayed ability to transmit these diseases may be explicable on a quantitative basis, the amount of virus that the insect can pass out into the plant increasing with time until a quantity sufficient to cause infection is reached; the length of time the vectors remain infective has been shown to depend directly on the period during which they had previously fed on the diseased plant.

In a second and much larger group of virus diseases, including most of those transmitted by aphides, the insect becomes infective immediately after feeding, and there is no evidence of multiplication of the virus in the vector. In such cases there is obviously a possibility that the insect serves as a merely mechanical instrument of transmission, similar to a needle. It has been found, however, that aphides that have had a preliminary period of fasting are more highly infective than those that are allowed to feed, and can transmit infection to a series of plants, while those that have fed freely usually fail to infect more than a single plant. It seems evident that the virus has to be taken into the insect even in this group. No evidence has been obtained so far that any plant virus undergoes a necessary part of its development in the insect, as the malarial parasite does in the mosquito.

The power to transmit a virus is not equally possessed by all individuals of the vector species: The most interesting cases are those in which ability and inability to transmit are inherited characters following a simple Mendelian rule, as in *Cicadulina mbila,* the vector of maize streak. Noneffective *C. mbila* may be made infective by releasing the gut contents into the blood by puncturing the intestinal wall after feeding on a diseased plant. A similar case is known in the mosquito vector, *Aëdes aegypti,* of the animal virus disease, equine encephalomyelitis in a few cases the virus cannot be taken up by adult insects but only while they are still nymphs while the exact opposite has been reported in the transmission of beet crinkle by *Piesma quadrata.* The virus of rice dwarf passes from the mother to her offspring in *Nephotettix apicalis,* so that this leaf hopper can be a vector even when it has never fed on a diseased plant.

Ability to act as a vector is clearly, not a simple matter. So also the relations between the virus and its host plant are complex. The production of local necrotic lesions instead of systemic invasion is well known amongst the hosts of many viruses; both types may sometimes be found in different varieties of a single host, such as the potato. These local necroses preventing systemic spread seem to be examples of hypersensitivity similar to that to which many plants owe immunity from rusts. Sometimes all attempts to inoculate by needle pricks fail, whereas the stylet of the vector succeeds; this may be due to the slight injury caused to the cells by the latter as compared with a needle, for the needle has been successfully used to inoculate the vector of maize streak but not the plant. It is possible that the enzymes present in the vector's saliva or in the plant cells of the host may playa significant part in the process of infection as they do in infection by parasitic fungi, but there is as yet little evidence of this. With some viruses only young leaves can be successfully infected by the insect in others there is a gradient of susceptibility between leaves at different heights on the plant, as in tobacco mosaic. It seems evident that the behaviour of viruses as infective agents is governed largely by physiological factors, and that, as with fungi, successful infection is a delicate process dependent on conditions affecting host, vector, and parasite.

Transmission by grafting succeeds in all cases in which full organic union can be secured between a diseased and healthy susceptible host. It is the only method known in a few cases, the number of which is decreasing as their insect vectors are discovered.

Transmission through seed is known in bean, clover, cucumber, lettuce, and other mosaic diseases. It cannot occur where the virus is restricted to the phloem, as there is no vascular connection from mother plant to embryo. In all cases of seed transmission that have been adequately investigated, the virus has also been found in the pollen. External contamination of the seed may be responsible for infection of the cotyledons and other parts of the germinating seedling, and from the practical point of view may be as important as true seed transmission. Of much greater practical importance, however, is the spread of infection by the use of diseased tubers, bulbs, roots, suckers, grafts, and the like in plants that are propagated vegetatively.

The movement of the virus within the plant may be restricted to the phloem as in sugar-beet curly top, or may be independent of the vascular tissues. Move in the phloem may be very fast, the curly top virus having been found 15 cm. down a sugar-beet seedling leaf 6 minutes after the vector was placed on the leaf, and maize streak virus 4° cm. down, 2 hours after feeding commenced. Tobacco mosaic virus, on the other hand, shows little movement for 2 to 4 days in super facial inoculations but then passes rapidly from the inoculated leaf towards the root and top of the plant; a distance of 7 inches was traversed in 1 hour in some tests on tomatoes. In the rapid movement, lateral organs such as leaves and fruit trusses may be missed for a time. It is believed that the virus on inoculation moves slowly from cell to cell through the plasmodesms until it reaches the phloem, in which it travels fast. In cowpeas inoculated with tobacco necrosis virus it took 12 days for movement to become rapid and 48 hours later the distance traversed was doubled. In passing across a leaf from one epidermis to the other, tobacco mosaic virus travelled only 7 or 8μ an hour. The rapid rate of virus movement through the phloem is in harmony with what is known regarding the movement of solutes through this tissue.

The incubation period of plant virus diseases, that is, the time elapsing between inoculating and the first appearance of symptoms, is very variable but is sometimes longer than that known in any fungal or bacterial disease~ The shortest period that has been recorded appears to be 18 hours for the tobacco necrosis virus. At the other extreme is apple mosaic on certain hosts, since 20 to 27 months elapsed before symptoms appeared in inoculated roses 6 inches below the point of inoculation. The earliest symptoms of 'yellows' in peaches infected by means of *Macropsis trimaculata* appeared in one set of recorded experiments in from 14 to 268 days, but when transmitted by budding the first symptoms may take 2 or 3 years to appear. In the' phony peach' disease the incubation period is about 18 months, and in 'sandalwood spike' 7 months under natural conditions and 4th months in artificial inoculations.

Some viruses are localised in particular parts of the plant. The most striking example of this is the virus causing 'phony' peach disease in the United States which is found only in the roots and is not transmissible by sap or any known insect vector. It has been successfully transmitted only by root-grafting; although a healthy scion grafted on a diseased stock will show symptoms, it will not infect another tree if grafted or budded into it. Though some toxic principle is carried up from the roots, causing serious dwarfing and crop reduction, this cannot be the virus itself or it would be transmissible by above-ground grafts. The tobacco necrosis virus is also present ordinarily only in the roots, though it sometimes becomes systemic and can be naturally or artificially inoculated into the leaves to produce local lesions. Several viruses occur mainly in the phloem, the sugar-beet curly top virus so much so that the other tissues seem to be actually toxic to it. A very few, of which the phony peach virus is the best example, are apparently restricted to the parenchyma, but the majority, including all the simple mosaics, are found both in phloem and parenchyma.

The chief vector of maize streak, *Cidcaulina mbila*, can pick up the virus and remain infective for 9 weeks after feeding for only IS seconds, much too short a time for the stylets to have reached the phloem where the insect normally feeds. Several viruses travel so slowly to distant parts of the plant that it may be a considerable time before all parts are reached. In sugarcane mosaic healthy tillers can be found at the base of diseased plants, and in potato leaf roll the harvesting of the tubers intended for seed while still immature has been advocated as a means of reducing loss from this disease.

INFLUENCE OF THE ENVIRONMENT ON VIRUSES AND VIRUS DISEASES

As in diseases caused by fungi or bacteria, temperature and other weather conditions may influence the incubation period, course, and effect of virus diseases. The viruses themselves are often more resistant to heat than are fungi, an extreme case being that of tobacco necrosis which has a thermal death point at 90° to 92° C. but can withstand dry heat at 100° C. for 15 minutes.

This virus is also very resistant to chemicals and retains virulence after 6 months in absolute alcohol, while the beet curly top virus resists I in 5° mercuric chloride, 1 in 25 carbolic acid, and I in 200 copper sulphate. Another resistant virus is that of ordinary tobacco mosaic which will stand 1° minutes at 93 ° C. though only I minute at 96°; it can also endure long exposure to mercuric chloride at germicidal concentrations.

In the cured leaf this virus required 1° hours at 100° C. dry heat for complete inactivation. Potato virus 'Y' is inactivated by 10 minutes at 52°C. and by 75 per cent alcohol, while tomato spotted wilt virus is even less resistant to heat. Ultraviolet light destroys viruses more slowly than bacteria or even bacterial spores.

In early experiments the incubation period of tobacco mosaic was shortest at 28° to 30° C., the development of symptoms being slowed down by lower temperatures proportionately to the growth of the plant; little or no symptoms appear at temperatures that check growth or if the plants are kept at high temperatures such as 36° or 37° for a fortnight. The incubation period for this disease was found in New Jersey to be 4 or 5 days in June and July as compared with 6 or 7 in December and January. Even severely infected mosaic tomato plants were noted as early as 1917 to have the symptoms completely masked at an average air temperature of 85° F. Similarly it was reported from Canada in 1921 that the symptoms and loss of potatoes affected by mosaic or leaf roll might be suppressed in the dry warm summer climate of the western areas, and this climatic effect has since been fully confirmed, e.g. in Kansas where mosaic, though completely masked, may seriously reduce the yield.

Light, soil temperature, and soil moisture have little effect on the foliage symptoms of potato virus diseases, but an air temperature of 25° C. is sufficient to mask most of them; yellow dwarf differs from the others tested in the United States in being very severe at 25° but. masked at IS°. The yellow edge disease of strawberries is difficult to diagnose at certain seasons, the symptoms being most evident in the south of England in late autumn, though severe cases can be recognised from June onwards; the temperature does not usually rise sufficiently for clear symptoms to appear before June, but even then the soil may be too dry to allow them to develop, since there seems to be a close correlation between symptom expression and soil moisture.

In the control of this disease by roguing it is necessary to take into account the weather during the previous week or two and select periods likely to allow symptoms to be clearly seen. The effect on virus diseases of the nutrition of the host has not been as extensively investigated' as in other types of disease. The information available, however, suggests that a balanced nutrition is as important in reducing the losses they cause as in those due to fungi or bacteria.

Susceptibility to yellow tobacco mosaic can be influenced by the level of supply of nitrogen, phosphorus, or potash, the phosphorus effect being related to the rate of growth of the host. With nitrogen or potash, however, some other factor seems to be concerned. The influence of nutrition on the concentration of the virus in the tissues is shown by the fact that high nitrogen feeding can increase the concentration of tobacco mosaic in the cells to about five times that in normally fed plants, while 8 days after inoculation the juice from plants receiving high doses of nitrogen may contain twelve times more virus than that from low nitrogen cultures.

The ecological relationships of the insect vectors are even more important in the incidence of virus diseases than the effects of climate n the diseases them selves. They have been most fully studied in the aphis vectors of potato virus diseases and the leaf hopper that transmits sugar-beet curly top. In North Wales cool, damp windy areas occur near the coast in which potatoes have been grown for long periods without degeneration.

The initial infection of the crop each year has been traced to the migration to it in early summer of winged Myzus persicae, the principal vector of the diseases causing degeneration, chiefly leaf roll in the area in question.' These aphides over-winter. in the wingless form on winter crops of savoy cabbage, kale, and other brassicas which, as already mentioned, may be symptomless carriers of leaf roll.

A certain number of the winged aphides migrating from brassicas to potato fields may contain the virus, and each serves as a focus of infection, while others settle on potatoes infected by way of the tubers from the previous crop and so pick up the virus. Migration of the winged form occurs almost exclusively when the temperature reaches 65° to 70° F., when the relative humidity is below 70 to 80 per cent., and the wind velocity is less than 3 or 4 miles an hour. Under these conditions the insects will readily find their way to potatoes from brassicas at least a quarter of a mile away and then move actively in the potato crop from plant to plant and field to field; the plant-to-plant spread within a field, however, is mainly effected later by the crawling wingless form. In North Wales this period of active- spread of leaf roll is usually between late June and the end of August.

Where degeneration is rapid, the infestation may reach 500 to 1,000 aphides per 100 potato leaves, whereas in the good seed areas not ~ore than 20 may be found. Counts over a number of years have shown that infestation varies greatly from year to year, and there is evidence that the percentage of virus infection in the potatoes may be correlated with the intensity of infestation by Myzus persicae in preceding years. The time of the flight of the aphides is also important, for a small amount of infestation brought in early in the season might be spread considerably by wingless forms, whereas later there is less time for this to happen before harvest.

Almost the same degrees of temperature, humidity, and wind govern the migration of winged forms of the strawberry aphis, Capitophorus fragariae, that transmits yellow edge and crinkle of stranguries in south-eastern England. This aphis is able to find and colonise even isolated strawberry plants at least 400 yards away. Its winged generation is active for short periods in May and June, when extensive spread of the viruses takes place.

Similarly with sugar-beet curly top in the western United States the virus, which has a host range of 75 species of plants in 48 genera of 18 families, is brought into the crop by the spring generation of the leaf hopper vector which migrates from the weeds and bushes of the arid uncultivated plains and foothills as they dry up.

Both the magnitude of the migration and the percentage of individuals that harbour the virus vary greatly from year to year. In one area it has been found that when November rain allowed early vegetation in the dry zone, the spring migrants had from 16 to 42 per cent of infection, whereas in seasons of late rains only 2 to 6 per cent carried the virus. In southern Idaho the percentage of infected spring-brood *Eutettix* was found to vary from 4 to 67 in different years; sometimes the virus overwinters in living leaf hoppers, sometimes in susceptible desert vegetation, sometimes in the cultivated distrzicts.

In eleven years there were six of relatively late leaf-hopper movement and high yields, whereas in the five years when migration was on or before the 24th May the yields were low because of early infection of the crop; by harvest time practically all the leaf hoppers on beet contain the virus.

It is scarcely necessary to mention that the area of distribution of the vector may determine the distribution of a virus disease. This is well illustrated by peach yellows which coincides roughly in the United States with its vector, the plum leaf hopper *Macropsis trimaculata.*

VIRUS STRAINS: VARIANTS

A character possessed by viruses which brings them closer to the organic than to the inorganic world is the occurrence within a single virus of strains differing from one another in physical characters, the symptoms they cause, and virulence. Strain differences have been most fully studied in the potato mosaic virus 'X' and that of ordinary tobacco mosaic but they are marked in cucumber, sugarcane, and many other viruses.

Experiments at Cambridge reported in 1933 showed the inoculation into tobacco of potato virus 'X' did not always result in the same symptoms. When the virus was taken from the green and the yellow areas and from necrotic rings respectively, on the leaf, three strains could be differentiated. About the same time the discovery was made in the United States that the yellow form of tobacco mosaic, previously thought to be due to a distinct virus, sometimes arises spontaneously in plants inoculated with the ordinary form and that ordinary tobacco mosaic contains many strains. Three of these strains cause the well-known mild, aucuba, and distorting mosaics of the tomato in England.

The term 'mutant' has been applied to strains appearing spontaneously or produced by heat, irradiation, and the like, but since there is doubt whether the viruses are organisms it is better to adopt the term 'variant' used by some authors.

The variants recognised by virus workers agree with mutants in being permanent departures from the parent type, though, like mutants, they can throw further variants or possibly revert to the parent type. Some of the variants of tobacco mosaic differ remarkably in the symptoms they cause ring and necrotic spots, filiform leaves, and in their virulence, rate of movement in the plant, and other properties. The same is true for potato virus 'X' its variants may appear spontaneously or be induced by passage through other hosts, sometimes being so weak as almost to become latent and without symptoms. The 'X' variants studied in England tended towards reduced virulence, but some of those found in Germany caused much more severe symptoms when inoculated into tobacco than the parent form.

By exposure to X-and gamma-rays, variants of the purified tobacco mosaic virus have been produced with distinct chemical and biological properties. The former seem to be due to changes in the nucleic acid part of the virus molecule, not greatly affecting the molecular weight but causing different solubility and hydration relationships, and in some cases an increased tendency to polymerisation.

The biological changes are marked not only by differences in symptoms in tobacco but by a delayed incubation period and sometimes an increase in the number of local lesions produced by a given quantity on leaves of *Datura stramonium* or *Nicotiana glutinosa*; this last may be due to an accretion of virulence or to multiplication of the virus particles, or to disaggregation of the purified rod-shaped paracrystals into the smaller units of which they are composed.

Unfortunately no way of proving that a single lesion is caused by a single particle of virus has been found, as can be done with single bacteria or fungal spores, and formal proof that all these types of, variant are due to processes akin to mutation or saltation, and not merely to the separation of two or more unit particles, is lacking though the presumptive evidence for it is strong.

RESISTANCE AND SUSCEPTIBILITY

The cultivation of resistant varieties of crop plants has not proved as yet to be such a potent weapon in the fight against virus diseases, with certain notable exceptions, as it has in combating diseases due to fungi. Natural resistance against potato leaf roll and tobacco mosaic is rare, and without field evidence of its occurrence in these old-standing diseases the search for resistance was long delayed.

In their efforts to control the destructive sugarcane virus disease 'sereh' in Java, the often-told story of which is one of the brightest chapters in phytopathological history, Dutch workers had a certain measure of early success in crosses between thick or 'noble' varieties of Saccharum officinarum and the thin Indian variety Chunnee of the species S. barberi. Some of these proved tolerant also of mosaic, yielding almost as heavily when infected as when healthy, and their use, especially the P.O.J. variety '213' introduced into cultivation there in 1917, saved the cane-growing industry in Tucuman, Argentine, from imminent extinction.

In high temperatures, however, these varieties tend to lose their tolerance. Later crosses of *S. officinarum* with Kassoer, a presumed hybrid between it and the mosaic-immune wild species *S. spontaneum,* yielded forms such as the P.O.J. '2,700' and '2,800' series which proved highly resistant to mosaic and in some cases to sereh. These rapidly spread throughout the tropics. In other countries also varieties resistant to mosaic were bred; in the West Indies crosses between the resistant Java canes and the rich susceptible canes bred in Barbados have proved that resistance amounting to field immunity occurs when only one-eighth or one-sixteenth of the 'wild blood' of *S. spontaneum* is present, and in itself the replacement of the older varieties by P.O.J. '2,878' reduced sereh and mosaic to practical insignificance before 1928. For a time other countries fought mosaic by growing the naturally resistant Uba and other *S. sinense* varieties, but by 1931 most of the world's supply of cane sugar was derived from varieties containing the Kassoer wild 'blood'. Several of the best of these varieties are also practically immune from a third virus disease, streak, in Natal, though curiously enough not in Egypt, while in Queensland their high susceptibility to Fiji disease and downy mildew has caused them to be barred where these diseases are prevalent.

The sugarcane mosaic virus does not rapidly nor easily become systemic; varieties may resist it or tolerate it and even recover so completely that it becomes impossible to transmit the disease to healthy canes. In Louisiana the varieties tested have been grouped into those that readily take and retain mosaic infection, those that take it with difficulty but retain it, and those that take it with difficulty and readily throw it off. A further complication is introduced by the occurrence of strains of the virus, some of which infect varieties that resist others.

Thus, a virulent yellow strain, first seen in 1933, caused heavy losses in practically all the commercially important varieties of cane then cultivated in Louisiana, including those that -had previously withstood the disease. Furthermore, some of the improved varieties tested in Louisiana have shown so great a susceptibility to fungal diseases that they have had to be discarded. These difficulties have led to the establishment in recent years of an elaborate system of testing varieties before release for cultivation, and illustrate in striking fashion how complicated crop breeding for disease resistance can be.

In the breeding of potatoes for virus resistance a notable advance has recently been made at the research station of the Scottish Society for Research in Plant Breeding at Corstorphine. It was found that certain varieties which react to infection by the ubiquitous 'X' virus by a lethal necrosis of the green shoots are ordinarily free from the disease in the field, presumably because infected plants are killed off. The same happens with the 'A' virus. King Edward and Epicure crops are thus freed from both these viruses, other varieties reacting in the same way to one or the other or to the less important 'B' and 'C' viruses. The breeding programme was adjusted to combine lethal reaction to all these in one variety, and already Craig's Defiance, which is intolerant of 'A', 'B', 'C' and 'X' has been produced. The analogy between this and the use of the hypersensitive reaction in breeding against cereaf rusts is clear. True resistance to the 'X' virus has been obtained in the United States in a seedling'S 41,956 while another seedling variety Katahdin shows field resistance to 'A' and to some extent to 'Y' but is susceptible to 'X'. Attempts are now in progress to combine resistance to 'X' and 'A' the two viruses important as constituents of mild mosaic in America.

Resistance to tobacco mosaic may be the result of hypersensitivity marked by the production of localised necrotic lesions as in the species *Nicotiana glutinosa,* or the virus may become systemic but causes few symptoms and little damage, as in certain Colombian varieties of *N. tabacum* of which Ambalema is the best known. Both these types have been used in breeding mosaic-resistant commercial tobaccos. The factor causing the local-lesion reaction has been transferred to *N. tabacum* type plants both in the United States and in Russia, but the most recent American report does not speak well of them. Crosses of Ambalema with good commercial types have been under test in the United States since about 1936 and are giving excellent promise as regards immunity, growth type, and quality. In Rhodesia, while Ambalema showed no trace of mosaic on inoculation, the progeny of some of its crosses with good susceptible varieties were extensively infected, though they outgrew the disease to a considerable extent.

The strawberry yellow-edge disease which is responsible for much of the deterioration of the British crop that has been marked in recent years, causes little obvious signs of infection in certain varieties, largely of Continental origin; most of these, however, are carriers of the virus. Of the two parents of the cultivated strawberry, *Fragaria virginiana* and *F. chiloensis*, the former is highly susceptible and shows the symptoms readily and continuously, while the latter is highly tolerant and usually behaves as a symptomless carrier; the commercial hybrids probably react largely according to the degree to which these two parents are represented. The woodland strawberry *F. vesca* resembles *F. virginiana* in its reaction and serves as a useful indicator plant for revealing masked infection. Breeding against yellow edge and the crinkle-virus disease which usually accompanies it cannot with safety, on present knowledge, be directed to the production of carrier varieties, for it has been reported that they may suffer progressive deterioration and eventually 'run out' like such fully susceptible varieties as Royal Sovereign. For the time being, there fore, control is being attempted by the development of virus-free clonal stocks of these fully susceptible varieties and their maintenance under nursery conditions by roguing and the control of the aphis vector, *Capitiphorus fragariae.*

In the United States the production of virus-resistant crops has reached the stage of commercial development in curly top resistant sugar-beet as a result of prolonged work by the Federal authorities. In 1937 the resistant U.S. '12' variety gave 12.6 tons per acre in test plots in Idaho against 0.64 ton for susceptible varieties, and this variety showed high resistance also in California in three-years' tests reported in 1938. Resistance to the same virus and to common bean mosaic has been developed in selections of the small red Mexican variety of *Phaseolus vulgaris* in Idaho so as to give field immunity from these diseases. In New Zealand a number of varieties of garden and field peas have shown high resistance to the common pea mosaic virus and their improvement is in progress.

Though the examples cited indicate that breeding for virus resistance is as yet in its infancy and shows more promise than achievement except in sugarcane, one of the most striking instances of control of a plant disease in the course of a very few years by the selection of resistant strains is afforded by the leaf-curl virus disease of cotton in the Sudan. Leaf curl was first seen in the Gezira in 1923, and within ten years was present in every field in the 200,000 acres of irrigated cotton in this area, often causing 100 per cent infection and at times almost complete loss of crop. Individual plants of the Sakel variety of Gossypium barbadense, chiefly grown in the Gezira, showed marked resistance to leaf curl; two selections from these, X '1,530' and X '1,730', which had only 3 and 2 per cent of leaf curl respectively, in a test in which the main crop showed 9 1.5 per cent., proved suitable for cultivation and most of the area was sown with them, especially the latter, within a few years. By 1938 leaf curl had been reduced to negligible proportions, though late attacks were still seen. Strains of the virus seem to exist, and there has been some recrudescence of severe infection possibly due to this cause in the last couple of years; there are now, however, no grounds for the fears expressed when the epidemic was at its height, that leaf curl would render uneconomic the cultivation of the valuable *G. barbadense cottons* in this important area.

The genetical bases of reaction to virus diseases have been determined in some plants. Separate dominant genes seem to be responsible for the lethal necrotic reaction to potato viruses 'A', 'B', 'C' and 'X' mentioned above; the resistance to 'X' of the U.S. Seedling '41,956' has been determined to be due to two necessary cumulative dominant factors. The resistance to tobacco mosaic of Colombian varieties of *Nicotiana tabacum,* including Ambalema, is due to two necessary recessive genes, and the local necrotic reaction of *N. glutinosa,* chillis, and *Browallia speciosa* to this virus depends on a dominant gene; modifying factors have been found to influence the results in most of the crosses studied. Resistance to lucerne mosaic in some of the varieties of *Phaseolus vulgaris* has been found to be due to duplicate dominant genes.

As mentioned in an earlier chapter, a method of inducing immunity in virus susceptible plants was reported from Java in 193 I; tobacco infected with the common green mosaic virus could not then become infected by the whitish-yellow strain, and vice versa. Independent investigations at Cambridge reported in 1933 showed that the same result followed the inoculation of tobacco with a green and a yellow strain of potato virus 'X'. Subsequent work has fully confirmed the existence of this kind of induced immunity and has also shown that plants which have recovered from virus attack may remain outwardly immune from reinfection with the same virus. 'Recovered' plants, however, may continue to harbour the virus in a latent or 'diluted' state though it is still fully active when tested on susceptible plants; the power to resist reinfection may sometimes be capable of transmission from plant to plant by grafting, and it has been suggested that this is an example of passive immunisation.

There is some evidence that the degree of protection given by one virus strain against a second related one is quantitatively proportional to the amount of the first virus present in the tissues, and it is suggested that in some way there is an intense competition within the plant between related viruses since it is only with these that acquired immunity is found. Sometimes the presence of a virus does not prevent the invasion of the tissues by a second related one but only masks or modifies its symptoms, so that there appears to exist a form of dominance of one strain over another; this may be complete or partial or may be temporary, being replaced after a time by mixed patterns.of the two types of symptom.

When the two strains are inoculated simultaneously, one may become dominant or the two may act in equilibrium, and a mixed pattern result; complete dominance is sometimes shown at the tip of tobacco plants inoculated with a mixture of two strains of ordinary tobacco mosaic, while lower down the symptoms of both are apparent on the same leaf. Recent work does not support the analogies that were earlier drawn between the immunity acquired by 'vaccinating' plants with weak virus strains and that conferred by vaccination against human and animal diseases; there is no evidence of antibody formation in the plant, and the observed facts point rather to a pre-emption by one strain of the sites of multiplication of the second. When the two viruses are not related, one does not prevent the multiplication of the second.

Nevertheless, viruses produce serological reactions when inoculated into animals and can induce the formation of antibodies in the animal's serum. These antibodies, like those resulting from the presence of animal pathogens, neutralise the virus which caused their production and also give the precipitin reaction by causing a precipitate to form with sap from infected but not from healthy plants. The reaction is ordinarily specific and given only with related viruses, e.g. those which have the power of inducing immunity against one another; a few exceptions, however, have been reported, as in potato viruses 'A' and 'Y' which are serologically related but do not confer immunity against each other, and differ in other ways.

Mere failure to immunise is by itself an unsafe criterion of specific difference, since it is well known that strains exist of human and animal viruses which, while causing identical symptoms, afford no immunity against one another. It has been found possible to separate allied strains of a given virus serologically, as they can produce specific strain antibodies in addition to those common to all strains of the virus. The serological reaction sometimes affords a useful method for field identification of virus diseases.

Another property which plant viruses share with many animal pathogens is that of becoming attenuated in virulence by passage through appropriate hosts. The sugar-beet curly top virus has been attenuated by passage through the nettle-leaved goosefoot and remained stable even after several generations in older sugar-beet leaves, but regained virulence in the chickweed and sometimes in sugar-beet cotyledons or early leaves. Attenuation of potato virus 'Y' has been observed in *Schizanthus retusus*, and also as a result of long cultivation in tobacco.

ANATOMY AND HISTOLOGY OF VIRUS DISEASES

There is often no apparent change in the tissues and cells of virus-infected plants. A good many viruses, however, produce necroses in various organs, some produce distortion or abnormal development of organs, with or without hyperplasy or hypoplasy, while intracellular disturbances are not uncommonly marked by inhibition of plastid formation or the presence of abnormal cell inclusions.

In many of the true mosaic diseases there is some hypoplasy in the tissues of the lighter green or yellow areas of the leaves, the palisade cells becoming short and almost isodiametric or even, in early infections, remaining undifferentiated from the rest of the mesophyll, and all the cells and intercellular spaces being smaller than usual so that the thickness of the leaf is reduced. In cotton leaf curl large intercellular spaces are found in the rectangular palisade tissue.

The necrosis of the leaf blade produced by viruses which cause local lesions has been studied at Rothamsted in Nicotiana glutinosa inoculated by the strain of tobacco mosaic responsible for aucuba mosaic of the tomato in England. It begins by the formation of a dark-staining band between the cells of the epidermis and spongy parenchyma or occasionally between those of the palisade tissue, and extends through the leaf all around the inoculated spot. Nuclear, but not cell, divisions occur in the area enclosed by the band, the cells die and dry out, and the dark-staining material, which is very insoluble and chemically refractory, appears to prevent the virus from,passing to cells outside the iesion.

Three types of internal necrosis of stems have been distinguished in certain virus diseases of the potato. In the 'acronecrosis' or killing of the top of the plant which may be caused by potato viruses 'A', 'B', 'C' or 'X' necroses develop especially towards the top of the stem, starting usually in the internal phloem and spreading to the adjacent parenchyma. A phellogen occasionally arises later in the parenchyma and tends to isolate the necrotic area by a ring of cork.

The necrotic process begins by a thickening and separation of the cell walls of the primary phloem elements, the intercellular spaces thus formed becoming filled with a yellowish-brown gum-like substance. Similar changes occur in the phloem parenchyma and often cause collapse by pressure of the sievetuhes and companion cells. The necrosis spreads to the xylem parenchyma and may even involve the vessels. Sometimes the outer phloem shows similar necrotic changes. The petioles are also involved, so that individual leaves are killed. In the tubers, necrotic patches may form like those in the stem, except that the cells of the storage parenchyma near the phloem are more extensively suberised and the necrotic areas always become eventually isolated by cork.

In the second type, the 'acropetal' necrosis caused by potato virus 'Y' the collenchyma in leaf veins, petioles, and stems is mainly affected, the vascular tissues remaining normal. Spread occurs vertically, causing streaking of the green parts, and also along the collenchyma around the stem. The cell walls become suberised and the cell content disappear or are replaced by gummy matter.

The third type is characteristic of potato leaf roll, and is restricted to the phenol. The walls of many of the primary internal and external phloem groups are thickened and lignified and small spaces appear between them, filled with 'a brown deposit. Eventually the phloem throughout the plant is affected, though the underground parts may be reached late; a visible' net necrosis' due to browning of the phloem strands of the tubers occurs in some varieties. In severe cases pressure may destroy the elements of the primary phloem. The first cells to show lignification are usually those near the fibres, though occasionally it starts at the centre of a phloem group.

Among other virus diseases affecting chiefly the phloem, 'curly top' of sugar-beet in the United States, Fiji disease of sugarcane, and 'bunchy top' of bananas in the East, show marked tissue abnormalities. In curly top, degeneration begins by hypertrophy, followed by death of some of the phloem parenchyma cells nearest the sieve-tubes. Cells further out then divide, and a thick walled hyperplastic mass of incipient sieve-tubes without sieve-plates, but sometimes with companion cells, results. New phloem elements are still formed by the cambium but they consist mostly of imperfect sieve-tubes with companion cells and parenchyma. This parenchyma, with the pericycle, enlarges and divides into a callus-like tissue which may push up into small warts on the under surface of the veins, while the sieve-tubes and companion cells die and collapse. In the roots of infected beets the chromatin first increases, then breaks down with the cytoplasm. Other cells show a great increase in calcium oxalate and leucoplasts.

The changes in the phloem in bunchy top begin in the same way, but the hypertrophy and hyperplasia extend to the pericyclic fibres and ground parenchyma, which are largely transformed into a small-celled tissue rich in chlorophyll and visible as characteristic dark-green stripes bounding the veins.

In Fiji disease the galls always show on the under side of the leaf, but whether they originate in the phloem or are a product of more generalised meristematic activity is uncertain. The principal feature is still a hyperplastic phloem with callus-like proliferating parenchyma; this gives rise to sieve- tu be~ and tracheids. The vein-thickening which precedes and accompanies the development of enations in leaf curl of cotton and tobacco results from an increase of the primary phloem and hyperplasia of the pericycle in which bundles with central xylem form.

In the witches' broom of potatoes in North America, elongation of the shoots is due in part to unusually long internodes, while in potato 'spindle tuber' the tubers are lengthened by elongation of their cells, and both in tubers and shoots the ratio of length to breadth of the cells is increased. The small galls sometimes found in stripe disease of narcissus and daffodils are due to hyperplasia of the epidermal and palisade cells or sometimes merely to their hypertrophy.

Enations develop usually, but not always, near the veins and surround a chlorotic area on the under surface of the leaf. They arise by a proliferation of the lower three or four cell layers, emerging as paired green leafy ridges or shallow cup-like protrusions. All the cells may undergo hyperplasy, until growth ceases. The fully formed enation in tobacco mosaic has two epidermal, a palisade, and four spongy mesophylliayers, the palisade being on the inner side in paired or cupshaped outgrowths and being often continued across the bottom of the cup, while the outer side of the enation resembles the under surface of the leaf.

On tomato the enations are similar but in a type associated with' fern leaf' in Italy, warts composed of two or three layers of hypertrophied epidermal cells and ridges due to proliferation of the mesophyll covered by a hypertrophied epidermis have been reported. No marked deviations from the tobacco type seem to have been recorded in the enations of cotton and tobacco leaf curl and pea mosaic.

In the 'big bud' disease of tomatoes the vascular system of all the above ground parts is stimulated. Interfascicular vascular tissue develops precociously in the stems; the flower stalk and the elongated segment between the calyx and the corolla become polystelic; and the phloem in the stem and inflorescence is much increased, the sieve-tubes being filled with a yellowish deposit. In the fruits the bundles are woody from intensification of lignification; annular and spiral vessels are replaced by pitted ones.

The leaf mesophyllioses differentiation and contains only a few small spaces; in the petals, however, a palisade layer may form. The profuse development of adventitious root initials may be accompanied by longitudinal stem fissures exposing the inner tissues. The stem and petiole thickening, which may cause a remarkably bloated appearance, is due mainly to abnormal development, by hyperplasia of pith cells close to the internal phloem, of a compact tissue of small cells with a few sieve-tubes and isolated tracheids.

The phyllody of the floral axis is extreme, even the ovary being sometimes replaced by small normal leaves, through the centre of which the axis may be carried by a process of median floral proliferation and terminate in a small shoot or a group of phylloid flower buds. The short, dichotomously branched apical shoot when thus formed may bear minute, hyaline papillae, thought to represent ovules. Beans infected by the virus show similar floral abnormalities.

The pallor or yellowing of green tissues found in so many virus diseases is due to inhibition of plastid formation and, though this seems to be less common, destruction of existing chloroplasts. The chlorophyll content is reduced from 43 to 87 per cent of the normal by different strains of tobacco mosaic virus, the yellow pigments carotene and xanthophyll being also proportionately reduced; in the yellow strains chlorophyllase is increased. That there is no obligate relation between chloroplast and virus is evident from the fact that viruses have been propagated for long periods in isolated root-tip cultures.

The most notable effect of some viruses upon the cell contents is the formation of the so-called 'X-bodies' or cell inclusions, which for a time were thought by many to be visible stages of the virus. The inclusion bodies are of two kinds, amorphous and crystalline; they occur most often in the epidermis and leaf hairs but are also found in the ground parenchyma and the phloem. The' amorphous X-bodies are often amoeboid, vacuolate, with mitochondria and oil drops, and give the usual protein reactions. They are formed from aggregations of small particles in the streaming cytoplasm of the cell.

The crystalline bodies or 'striate material' may occur in the same cells or independently. They are plate-shaped, fragile, colourless, and give protein reactions; the slightest pressure causes them to break down into needle-like fibres. They may represent condensates of the virus protein, while the X-bodies are-more probably reactions of the cell cytoplasm. Small plate-like inclusions have also been found in the nuclei of solanaceous plants infected with the' severe etch tobacco' virus; these are more stable than the cytoplasmic inclusions. Isometric deeply staining crystals, 0. 3 to 4 μ in diametre, occur in the nuclei as well as the cytoplasm of broad beans infected with certain pea and bean viruses and are said closely to resemble those of severe etch.

LOSSES CAUSED BY VIRUS DISEASES

Though the damage caused by potato 'curl' was widely recognised a hundred and fifty years ago, many virus diseases are so insidious in their onset that the injury they do may not be appreciated or, more often, is attributed to other causes. There are, however, some virus diseases that are invariably fatal: 'peach yellows' always kills the affected trees; a 'spiked' sandalwood tree dies some fifteen months to two years after the first symptoms are seen; 'rosetted' peaches nearly always die the following autumn or winter; while the potato varieties that react by 'top necrosis' to certain viruses are usually killed off before they produce tubers. Several viruses cause complete loss of crop without killing the plant, as in 'groundnut rosette.' and bunchy top of banans. From these obvious extremes of damage, however, to the large group of viruses that cause only mild symptoms or even, in carrier varieties, none at all, every gradation can be found in the severity of the resulting injury.

The annual toll taken by virus diseases from the potato crop of England and Wales can be assessed with considerable accuracy at not less than £2,000,000 in normal times. In the United States the reduction in the yield of sugar beet from the 763,000 acres under the crop is reckoned to have averaged over a number of years, about 2,000,000 tons annually. 'Beet yellows' is estimated to have caused losses in the sugar and fodder beet crops in Belgium of over 60,000,000 fr. in 1937. Louisiana is calculated to have lost sugar to the value of over 100,000,000 dollars from sugarcane mosaic in the ten years from 1922, though at the beginning of that period there was still local doubt whether even 100 per cent infection did much harm. In the Argentine, mosaic caused a progressive reduction in sugarcane yields in Tucaman from the beginning of the century, reaching its climax in 1915 when the average yield for the previous five years was only 9 tons per acre. The introduction of Java mosaic-tolerant canes such as P.O.J. 36 and 213 giving yields of 20 to 30 tons per acre, saved the industry however, and by 1926, when they had completely replaced the older varieties, the province had the largest sugar yield in its history.

Many virus diseases of market garden crops and fruit also cause heavy financial loss. 'Lettuce yellows', due to infection by the 'aster yellows' virus, was recently estimated to have reduced the lettuce crop of New York State by an annual average of 5 or 6 per cent from 192 I to 1938; bean mosaic caused about the same loss of crop in Iowa from 1924 to 1926; cucumber mosaic 7 to 15 per cent in Iowa from 1924 to 1926; and 10 and 15 per cent., respectively, in Kansas and New York in 1928. The average loss from tomato mosaic in Utah in 1925 was 8 per cent.

Naturally much higher individual losses than these average figures occur and many have been determined in experimental tests. Sugarcane mosaic has caused a loss of So per cent in Hawaii and over 60 per cent in India.

In tests in Ireland leaf roll reduced the yield of British Queen potatoes from 9 to 3.7 tons per acre and of Ally from 11.4 to 2′7 tons.

When tubers from leaf-roll plants were used as seed for the next crop in Wales, the losses for three consecutive years were 55.S, 45.6, and 52.6 per cent respectively, showing that this disease can produce its maximum effect in the year after infection; very similar figures were obtained in a three-years' test in England.

In the United States lettuce yellows can cause 7° or So per cent of a crop to be lost and the spotted wilt virus can destroy three quarters of the yield; soy beans can lose 3° to 75 per cent of the beans and ordinary beans 5° per cent from the mosaics of these crops. The rapid spread of bunchy top of bananas in New South Wales led to a falling off of the yield from 65°,000 bunches from 4,75° acres in 1922 to 91,144 bunches from 1,002 acres in 1925, and many growers were ruined.

Early infections of tobacco mosaic can reduce the yield in North Carolina by 3° per cent and the value by up to 5° per cent and infection of tomato mosaic in Maryland can reduce the crop by over 5° per cent. In glass-house experiments in England early infections of tomatoes with the streak virus and tobacco mosaic virus caused yield reductions exceeding 20 per cent. Early infections with groundnut-rosette virus cause total loss; for no nuts develop after infection. In a test in Canada, raspberry mosaic reduced the fruit yield by 2S per cent by weight.

Even such losses as these by no means exhaust the destructive possibilities of virus diseases. The record of potato varieties in Great Britain and of sugarcanes in Queensland, the West Indies, and elsewhere shows many instances of the disappearance from cultivation of very valuable types owing to susceptibility to some virus disease. 'Running out' of the best varieties of strawberries and raspberries from the same cause is a present danger in Great Britain and North America. Even the tolerance of infection found in carriers may do no more than disguise this deterioration; the 30 per cent loss from potato virus 'X' in certain carrier varieties in Australia was unsuspected until revealed by tests, for the crops were mostly uniformly infected, while carrier varieties of strawberries may at times suffer as great a deterioration from latent yellow-edge virus infection, as the highly sensitive Royal Sovereign crop.

Another possible additional source of loss from virus diseases is the increased susceptibility of virus-infected plants to bacterial and fungal diseases. While not as evident as the enhanced susceptibility to certain human diseases caused by the influenza virus, there is a rather widespread belief that potatoes are rendered more susceptible to blight by becoming infected by viruses, and in Brazil a similar relation between tomato viruses and Septoria Zycopersici has been claimed.

CONTROL

A direct attack on virus diseases by the chemical methods so extensively used against parasitic fungi is out of the question. Viruses cannot be reached by these means and, at the most, control of their insect vectors may be undertaken when the economic conditions permit. - Practical control of the aphis vector of strawberry yellow edge and crinkle has been achieved in some cases in England by the use of vapourised nicotine. In commercial crops of the biennial drug plant, henbane a considerable reduction of the vector, Myzus persicae, was effected by weekly spraying with nicotine and soft soap during the first two months of vegetation in the first yeat. This gave a 3° per cent increase in yield in the third crop, harvested during May of the second year.

Roguing for the elimination of virus-diseased plants has had a considerable vogue, especially in crops propagated vegetatively such as potato, sugarcane strawberries, bananas, and the like; it is still practised extensively where the crop is valuable and the amount of infection moderate. Its principal use nowadays, however, is in the maintenance of 'foundation' or nursery stocks of disease-free plants which can be used for the planting of the main field crop. Where insect vectors can be kept down fairly readily and the plants are under observation closely enough for early infections to be recognised, as in glasshouse cultivations, roguing is of much value, as it is also, obviously, where the virus is seed-borne or the crop is propagated vegetatively and a healthy parentage is, therefore, essential.

Where the infection arises anew each year in an annual crop, success in reducing the source of infection by the destruction of 'volunteer' plants or out-of-season plantings of the crop, which may serve to carry-over the disease between successive main crops, has sometimes been considerable. Much benefit resulted from eradicating-volunteer cotton in the Sudan to reduce early attacks of leaf curl. In Southern Rhodesia a notice under the Tobacco Pest Suppression Act, 1933, prescribed the 1st August as the date by which destruction of all growing tobacco plants is to be completed as a control measure against tobacco leaf curl; this close season appears to cause the vector to die out. In California the enforcement of a celery-free period in districts where western celery mosaic causes severe damage has been effective in raising the yield, which had fallen from over 1,000 half-crates per acre in 1930 to 31 in 1934, to an average of 1,100 in 1939 and 1,000 in 1940. The growing of maize in the neighbourhood of sugarcane fields has been discouraged for similar reasons as it is susceptible to sugarcane mosaic and a favourite food plant of the insect vectors; this ban was enforced by legislative action in certain areas in Barbados in 1927.

Sometimes the eradication of weed hosts of the virus or of its insect vector has given good results. Destruction of the Russian thistle, *Salsola kali-tenuifolia*, on which the beet curly-top vector, *Eutettix tenellus*, and the virus are largely maintained before passing to the crop, has effectively reduced sugar-beet curly top in California. Excellent results in controlling the yellowish-green mosaic of cucumbers in the United States, where it has - many hosts, have followed the eradication of certain weeds, especially perennials. In England, where, however, this mosaic is less common in glass-houses than the mild or green-mottle mosaic which seems to be confined to the *Cucurbitaceae*, the houses in which it occurs should be kept free from susceptible flower crops and the common bryony should not remain near cucumbers or marrows. With the great prevalence of the spotted wilt virus amongst ornamentals, keeping the latter out of houses containing tomato seedlings is most important, while glass-house control in tomatoes and flower-crops alike must include elimination of the thrips vector from an infected house before another susceptible crop is grown; thrips control out of doors, however, is difficult.

The complete eradication of these diseases by destruction of all infected plants is being effectively carried out by large-scale campalgn Saffiongst perennial plants. It was early advocated for the control of peach yellows and rosette in the United States and later was recommended against sandalwood spike, together with the establishment of a sandal wood-free belt round the infected area in South India. Recently, powers have been taken in the Gold Coast to remove plants such as cacao from declared areas in order to check the spread of diseases, such as the swollen shoot virus disease of cacao, to uninfected areas.

The object is to establish zones free from all cacao around the infected areas. The Psorosis Act of 1927 provides for the destruction, without compensation, of any citrus tree infected by psorosis within the Union of South Africa, together with quarantine measures against new introductions of the disease; as infection may remain latent in the tree for 20 to 30 years and seldom shows before the trees are 9 or 10 years old, it will be long before the effectiveness of the measure can be judged. In 1939-40, however, only 1,468 affected trees were found in a total of 513,123 inspected.

When peach mosaic was first seen in Colorado in 1934, and California a year later, eradication was promptly undertaken; in Colorado only 3,100 fresh cases were found in 1937, compared with 9,835 in 1936 and 30,467 in 1935, while in the four years from 1936 to 1939, 89,355 cases were found in California, 63,651 of which were removed. The attempt to eradicate the phony peach disease from the south-eastern States is more formidable; during 1936 over 2,000,000 abandoned and 3,500,000 'escaped' peach trees were removed in II States by a force of 2,000 men employed under the Emergency Relief measures, while altogether over 21,000,000 trees were inspected in 20 States and 146,072 of the 156,977 found infected were removed. In New Zealand an attempt is being made to eradicate 'onion yellow dwarf' disease which was found to be confined to a small area In 1939.

The progress that has been made in the breeding of crop varieties resistant to, or tolerant of, virus disease has already been discussed. In the opinion of many this remains the most hopeful line of attack on them. Control by 'vaccination' with a weak strain of the virus concerned, so as to induce immunity from more virulent strains, has been regarded as practicable by some. No satisfactory case of its use under field conditions seems to have been reported as yet and much further testing is necessary before the many risks associated with the use of such 'live' vaccines can be discounted.

A considerable measure of success has sometimes followed the adjustment of agronomic practices so as to reduce the prospect of infection. The most effective has been the alteration of the date of planting the crop so that it is too far advanced for serious injury by the time the insect vectors become numerous; this has worked well on occasion against the curly top of sugar-beet. In Tanganyika, experiments show that cassava planted in June is less injured by mosaic than if planted at other times, for the crop then passes the main period of its growth when the probability of infection is at its lowest. Similarly in parts of Nigeria the native tobacco plantings escape serious injury from leaf curl because the vector has become scarce by the time the seedlings are transplanted.

Late sown oats and other cereals in Siberia are not severely injured by the Zakooklivanie or 'pupation' virus disease, because the vector, *Delphacodes striatella*, which over-winters in the larval stage on grass weeds about the cereal fields, migrates in the winged form too early to be able to dwarf the plants. Infection by the groundnut rosette virus in Africa has been found to be considerably lessened by dense planting and also where grass mulching of the soil is practised; in the

Gambia earlier sowings are less damaged than those that give young plants when the vector, Aphis laburni, is most active in July and August. Possibly, as with the vectors of leaf roll of potatoes, the activity of this aphis is reduced by high humidity.

The chief measure employed for the control of virus diseases of the potato and some other crops in the British Isles and elsewhere is the inspection and certification of planting stock. For this purpose an elaborate organisation has been built up by various Government Agricultural Departments. In areas in the British Isles where potato seed production is important, the inspection covers purity to type, as well as the amount of virus infection. In Scotland, for instance, over 50,000 acres grown for seed are inspected annually and certificates granted by the Department of Agriculture of different designations according to purity and the percentage of virus diseases found: 'Stock Seed' must be 99.5 per cent true to type and practically free from virus symptoms except negligible mottle; T.S., is 99.5 per cent pure and with not more than 1 per cent total mild mosaic, severe mosaic, leaf roll, and 'wildings', and comprises varieties immune from wart disease, the corresponding non-immune grade being designated 'N.I.(A)'; 'T.S.(H)' and 'N.I.(H)' are 99.5 per cent pure and with not more than 3 per cent severe mosaic, leaf roll, and 'wildings'.

These are all regarded in England as Class I Scotch seed; Class 1 Irish, and Class 1 Special English and Welsh stocks are also recognised, the latter from certain limited areas only. Recently certification has been extended to cover English crops which, though ordinarily grown for consumption, can, under certain circumstances, serve as sources of good seed; such crops are eligible only if grown from Class 1 seed and at least 5° yards from any crop not similarly grown. In Canada and Holland also there are important seed potato industries and a very elaborate certifying organistation exists; in 1935 the acreage certified in Canada was 83,537. Many other countries have adopted similar measures.

Methods for the same purpose in the United States vary to some extent from State to State but are kept in general consonance by periodical conferences at which Canada is represented. Considerable use is made of the tuber unit method of establishing clean stock for planting the main crop: a portion of each tuber bearing an eye is tested under glass before planting time and only if it gives a healthy shoot is the rest of the tuber used. The resulting seed plot is grown in isolation and carefully rogued to eliminate any virus diseases that may appear. Naturally, localities are selected for seed production where the risk of spread of virus diseases is low. By this method in commercial practice cases have been recorded in New York State of the reduction of virus diseases from 2,6 per cent in 1924 to 0'4 per cent in 1932, and in Quebec of an increase in yield from 287'7 bushels per acre in 1928 to 417 in 1934.

The outstanding importance of the ecological conditions controlling the insect vectors of these diseases is sufficiently domesticated by the fact that there are areas in Ireland and elsewhere in which potato varieties have been maintained free from degeneration for many years, Early Rose potatoes have been grown on bog-land in central Ireland since 1867 without change of seed and without running out, and almost forgotten varieties such as Pink Eyes and Lumpers on the western seaboard for much longer periods. In these areas the insect vectors are rare and the cool moist climate is adverse to their dispersal.

The bunchy-top disease of bananas in New South Wales and Queensland has been brought under fairly effective control by a series of measures directed to the eradication of diseased plants and the prevention of spread. All plantations are registered and inspected diseased plants are removed after treatment by prescribed methods to destroy the vectors present on them, disease areas are quarantined so that no banana material may be moved from them, in the areas must free from weeds for no an 6 feet from each banana plant, all banana plants in other than registered plantations must be destroyed, and every care must be taken in the plantations to recognise the early symptoms of infection. Permission to sell planting material must be obtained, and is granted only to growers of disease-free stocks. In certain areas free from the disease no planting material from the infected areas is allowed ingress, These stringent measures, applied under legislative authority in 1927 and varied from time to time, have been effective in restoring the New South Wales banana-growing industry-probably infected from Fiji in 1913 and which had fallen between 1922 and 1925 to a seventh of its previous production, back to far more than the old figures by 1935 - and rank amongst the best examples of plant-disease control by administrative action based on sound scientific investigation. Similar methods are applicable to other virus diseases in advanced countries, especially in so far as they relate to the inspection and certification of nurseries supplying planting material.

In New York State, for instance, all raspberry nursery stock offered for sale must be regularly inspected, rogued when necessary, and certified reasonably free from virus diseases. A test of the same procedure in England has broken down owing to the prevalence of carrier varieties and others with a low expression of symptoms.

From this survey it will appear that the question whether the viruses are, animal, vegetable, or mineral' is largely of academic interest only. As agents of disease they behave sufficiently like the bacteria to allow of the application of the same considerations to the study of their behaviour and control.

In particular, the problems of their origin, dissemination, and prevention, which at one time seemed so difficult as to lead many plant pathologists almost to despair of their solution, have yielded to intensive research sufficiently to give good grounds for the hope that the practical outcome will eventually equal that achieved in the fight against parasitic fungi and bacteria.

SOME DISEASES OF VIRUS

It attracted first attention in the United States in the Connecticut River valley and has since been recorded in many tobacco-growing areas throughout the world. Clinton in Connecticut was the first to show that the virus infects tomato, and the disease is now recognized as one of- the most common on that crop. Allard transmitted the virus to several species of the Solanaceae including pepper, petunia, henbane, jimson weed, black nightshade, ground cherry, and several species of *Nicotiana.*

Grant explored the host range of the virus and found 29 susceptible species in 14 widely separated families outside of the Solanaceae. The list includes beet, spinach, phlox, foxglove, zinnia; buckwheat, snapdragon, white mustard, turnip, Kenilworth ivy, and tassel flower. Holmes extended the host range further.

SYMPTOMS

In young tobacco plants the first visible symptom is a downward curling and distortion of the youngest leaves accompanied by chlorosis. As the leaves enlarge, abnormally dark green spots appear which develop into irregular crumpled blister-like areas while the rerpainder of the tissue becomes more and more chlorotic. Various degrees of leaf malformation follow, and some leaves show only a mild diffuse mottle.

As new leaves unfold, they continue to show the typical range of symptoms. In pink-flowered forms the normal colour pattern of the flower is disturbed by interspersion of light-coloured areas or complete suppression of colour. Old leaves may show no mottle but may contain large yellowish blotches or scattered necrotic spots, symptoms often referred to by growers as "rust." Plants are stunted in various degrees. With some strains of the virus extensive necrosis occurs, and with others the chlorotic areas are more intensely yellow and more extensive in size. The usual symptoms on tobacco are very similar to those on the same host incited by typical cucumber-mosaicvirus, and the two diseases on tobacco are often confused.

The typical appearance in tomato is a mild mottle of the foliage. With some strains of the virus yellow chlorosis is predominant. In other cases severe strains cause necrosis of leaf and stem and necrosis and distortion of the fruits. This phase is commonly called single-virus streak. Quite similar streak symptoms are incited by the synergistic effect of latent mosaic:virus of,potato and any strain of the tobacco-mosaic virus or by the former in association with typical cucumber mosaic.

THE VIRUS

- Typical-tobacco-mosaic virus.
- Tobacco virus 1.
- *Nicotiana* virus 1 K. M. Sm.

Marmor tabaci Holmes: The virus is distinctive in that it remains active in extracted host plant juice up to 25 years, longer than any other known virus. It is infectious in dilutions up to 1-1,000,000 and withstands heating up' to about 90°C. for 10 min. The virus protein was first described by Stanley in 1935. An immense amount of fundamental work on the physicochemical properties of the virus has been carried out since. For a summary of these studies refer to Bawden and to Stanley.

McKinney in 1926 first isolated a yellow-mosaic strain from tissue taken from yellow spots in normally affected leaves. He later suggested that such strains arose by mutation, and further work by others tends to confirm this view. Jensen later described strains from a total of isolated which varied in symptoms incited.

Many investigators since Iwanowski have studied,the cell inclusions characteristic of tissue infected with the tobacco-mosaic virus. Two types of intracellular inclusions are found. The first of these, often referred to as X bodies, are amorphous, protoplasmic, more or less vacuolate bodies which give positive protein reactions.

Attempts to remove these bodies by micromanipulative methods have not been successful. The striate material has some resemblance to the purified virus-protein crystals, and it sometimes appears as the X bodies disappear. It is not likely that crystalline inclusions are deposits of the pure virus, more probably they are combinations of the latter with some other constituents of the host cell

DISEASE CYCLE

Sources of primary inoculum consist of debris of infected plants which' remains in the soil, tobacco refuse from warehouses, and certain infected manufactured tobacco products contaminating hands of workers. Johnson in a series of tests found 67 per cent of cigars, 81 per cent of cigarettes, and 62 per cent of pipe tobaccos to carry the virus. Manufactured chewing tobacco and snuff carried little or no virus, probably on account of the high temperatures to which the tobacco is exposed in manufacturing.

Dissemination from plant to plant takes place through mechanical transmission. One means is through handling of tobacco and tomato plants,at transplanting. Other operations which afford ready means of transmission are suckering of tobacco plants and suckering and training of tomatoes to a single stem in hothouse' production of the latter crop. Contact by man, animals and machinery during cultivation of the crops is another means of spread of the virus. The virus is not known to be transmitted by insects from tobacco to tobacco. However, the false potato aphid, Myzus pseudosolani Theob., transmits the virus readily from tomato to tomato and tobacco.

VARIETAL RESISTANCE

Two types of resistance in tobacco are known. One type is that found in the variety discovered by Nolla in Colombia, in which the virus increases but at a slow rate and symptoms are very mild.. The other type is the local necrotic reaction found in Nicotiana glutinosa L. This character is dominant in the Fl hybrid of N. glutinosa X N. tabacum L., but since the plants are self-sterile, special means must be employed to transfer the necrotic character to *N. tabacum*.

CONTROL

The first precaution in control is to avoid infested soil and to avoid the use of tobacco refuse on land on which tobacco Dr. tomato is to be grown. Seedbeds should be steam sterilized and well removed room tobacco warehouses and from greenhouses in which tomatoes are growing. Care should always be exercised to avoid contamination of hands with the virus from tobacco products and above all to avoid handling potatoes or cucumbers before staking or suckering tomatoes or tobacco. Thorough washing with soapy water is an effective practical means of cleansing hands between operations.

Some progress has been made in the direction of breeding of resistant varieties of tobacco and tomato, and some commercial mosaic-resistant varieties of tobacco are in use. Valleau in Kentucky reported successful use of several commercial varieties bearing the *N. glutinosa* type of resistance.

CUCUMBER MOSAIC

A mosaic of cucurbit plants was reported in Ohio, Massachusetts, and Connecticut during the first decade of the cuiterit century, but delineation of a distinct infectious disease was-not recorded until 1916. The relatively rapid increase of the disease in 'many area of the United States after the latter: date brought it into the position of major importance on cucumber where it has remained up to the present. The first through comprehensive study of the disease was published by Doolittle in 1920.

While cucurbit mosaic was at first regarded as closely confined to vine crops, the virus is now known to consist of many strains varying in host range. Anextensive list of dicotyledons and some monocotyledons is now included, among the suspects of one or another strain of the virus. Among the diseases of economic crops incited by cucumber-mosaic virus, in addition to the mosaic of vine crops, are southern celery blight, spinach mosaic heart rot of banana, some forms of delphinium mosaic, pansy flower breaking, tomato mosaic, tobacco mosaic, string-bean mosaic, lima-bean mosaic, crucifer mosaic, pea mosaic, and beet mosaic. Many common weeds are suspects. The disease is world-wide in occurrence.

There are certain other viruses which may infect cucurbits, arid the resultant symptoms may be confused with those incited by the typical cucumber-mosaic virus. Watermelon is not a host of the latter virus as a rule, but it and other cucurbits may be infected by other distinct viruses, such as cucumber viruses 3 and 4 and the tobacco-ring-spot virus.

Rader et described a mosaic of muskmelon in New York, the virus of which is regarded as distinct from the typical-cucumber-mosaic virus. The squash mosaic-of California is also distinct, the virus showing some resemblance to that of tobacco ring spot.

SYMPTOMS

The signs of infection on cucumber are often variable and difficult to define. The mosaic symptom is commonly seen on young leaves near the growing tip, but in such leaves, as they grow larger, the pattern often shifts to a diffuse, almost indistinguishable, mottle. V-shaped necrotic areas may appear on older leaves, extending from the margin towards the midrib. A dwarfing of internodes of the vine becomes progressively more 'conspicuous as mosaic symptoms of the leaf recede.

Young fruits show symptoms ranging from a mild mottle to extensive warty malformation and often to an almost entire lack of colour, a phase referred to as "white pickle". Infrequently, affected plants show a distinct wilt. After midseason, vines may show death of all leaves except for a few near the growing point. Many such plants die prematurely and of course are of little or no productive value after midseason.

THE VIRUS

- *Cucumber-mosaic* virus.
- *Cucumber virus* 1.
- *Cucumis* virus 1 K. M. Sm.
- *Marmor cucumeris* var. vulgare Holmes.

The typical-cucumber-mosaic virus is now known to have many variants, which differ chiefly in the type of symptoms on certain hosts and in host range. They are very much alike in physical properties. The thermal inactivation point is 65 to 75°C.; aging in vitro, 6 to 19 days; inactivation by dilution, 1-1,000 to 1-20,000. A strain reported by Price incites only local lesions on zinnia, while other strains are systemic on that host. Price's strain is commonly used in cross-protection tests to determine, whether a given strain is a member of cucumber virus 1 group. Most isolates do not affect legumes and crucifers, but strains which infect snap bean, lima bean, garden pea, and cruciferous crops are known. Most strains incite only local lesions on cowpea, but the snap bean strain is systemic on that plant. The southern celery blight described originally as celery virus 1 has been shown to be a member of the cucumber virus 1 group. It has as wide a host range as any of the isolates described, infecting many dicotyledons as well as such important monocotyledons as corn and banana.

Ainsworth distinguished two closely related strains as cucumber virus 1 and cucumber virus 2; they are identical with or very closely related to the typical strain described by Doolittle. He also described two other viruses as cucumber virus 3 and cucumber virus 4; these are nontypical in that the properties are different and the host range restricted to cucurbits. The virus proteins of the last two viruses have been purified. The tobacco-ring-spot virus also infects cucurbits, and whereas the typical cucumber-mosaic virus does not infect watermelon, the former virus incites a severe disease in that host. Squash mosaic, as defined by Freitag, Kendrick, Takahashi, and *Middleton et al.* in California, resembles the ring-spot disease; the two viruses have rather similar properties, and the proteins of each appear as spherical particles under the electron microscope. Takahashi secured squash-mosaic-virus crystals from a purified preparation of the spherical particles.

DISEASE CYCLE

Doolittle and Walker concluded that the typical cucumber-mosaic virus was rarely transmitted, if at all, with seed of cucumber, muskmelon, squash, and pumpkin. They did find wild cucumber seed, however, a common means of overwintering of the virus.

Bewley and Corbett in England considered that the virus was transmitted in seeds of hothouse varieties of cucumber.

The New York muskmelon-mosaic virus described by *Rader et al.* was transmitted in muskmelon seed with rather high frequency. Kendrick and Middleton found the squash-mosaic virus to be borne in a low percentage of seed of muskmelon and squash.

Perennial host plants are the most common between-crop reservoir of the virus. In northern states milkweed, Physalis spp., and pokeweed are common weed hosts. In Florida, Doolittle and -Wellman found the wild wandering Jew the chief inter-season host of the southern-celery-mosaic strain. Current-season spread is chiefly by aphids. The melon aphid and the peach aphid are both effective vectors, while both the spotted and striped- cucumber beetles also transmit the virus. When cucumbers are picked frequently for pickles, mechanical transmission may become an important factor.

CONTROL

In greenhouse culture of cucumber, screening of ventilators to exclude aphids and frequent fumigation ordinarily control the disease. The eradication of wild hosts was studied extensively as a possible control measure and found to be feasible by Doolittle and Walker. The procedure is economically sound, however, only in very special cases, e.g., with celery in Florida.

The development of resistant varieties of cucumber is the most practical solution of control. Highly resistant varieties suitable as breeding parents are found in certain oriental varieties such as Chinese Long and Tokyo Long. Suitable resistant varieties are being introduced.

POTATO MOSAICS

Potato mosaic began to emerge from the complex known as potato degeneration disease about 1914 when it was described by Orton as a malady present both in the United States and in Europe. Beginning about 1920 it became apparent to investigators of potato viruses that more than one type of mosaic occurred. There are now known to be several distinct viruses of potato which incite mosaic symptoms. Much confusion arose in the literature, however, owing to a number of influential factors which were only gradually realised and evaluated. The important variables may be listed tersely as follows: (a) Each virus concerned consists of a number of strains which differ in the symptoms they incite on the varieties which they are capable of infecting. (b) Each host variety, being a vegetative clone, has a high degree of phenotypic uniformity but may respond differently from each other variety to one or another virus group, the reactions varying from complete immunity through symptomless carriage, systemic necrosis, and various degrees of visible symptoms. (c) Environmental factors" such as temperature and light, influence the expression of a given virus-host interaction over a range extending from complete masking to severe stunting, mottling" and necrosis. The final exact diagnosis of a mosaic disease, therefore, very often cannot be based on symptoms but must rest upon the physical properties of the virus extract, serological tests, cross-protection tests, and,upon standardized hosts' reactions under specific environmental conditions.

Among the mosaic diseases of potato three are widespread and rank high in importance.

They are designated herein as,

- Latent mosaic;
- Mild mosaic, or crinkle mosaic;
- Rugose mosaic.
- Symptoms. Latent Mosaic.

Standard varieties of potato which had been in culture in the United States for some years previous to 1920 are symptomless carriers of the latent-mosaic virus. This is true of many European varieties as well. Some varieties used in Europe, when infected, respond with an acute systemic reaction, known commonly as top necrosis, which may take the form of killing.

SYMPTOMS

- *Latent Mosaic*: Standard varieties of potato which had been in culture in the United States for some years previous to 1920 are symptomless carriers of the latent-mosaic virus. This is true of many European varieties as well. Some varieties used in Europe, when infected, respond with an acute systemic reaction, known commonly as top necrosis, which may take the form of killing, beginning at the tip of the main stem or of a branch and proceeding down the top until the plant is killed. Tuber production may be prevented in top-necrotic varieties, and if small tubers are formed,,they commonly show- internal necrosis. Such, a variety is essentially a non-carrier and is, for practical purposes, resistant. Still other varieties show chronic mottle, stunting, necrotic spots on leaves, and in some cases necrotic spots in the tuber. Only these, therefore, exhibit mosaic symptoms. Some varieties show a great deal of variation between individuals; this is particularly true in newly introduced varieties and will be discussed more fully later.

- *Mild Mosaic.* The disease is incited by one or another of the strains of the mild-mosaic virus, and in the case of symptomless carriers of the latent-mosaic virus the synergistic effect of the latter may be a factor in disease development. This disease has also been referred to in America as crinkle mosaic. Schultz and Folsom at one time regarded mild mosaic and crinkle mosaic as distinct diseases. Dykstra in a later comparative study considered them to be incited by related strains of the same virus. Many old American varieties are susceptible to this disease, e.g., Bliss Triumph, Green Mountain, Early Rose, Early Ohio. Others, such as Rural New Yorker and Russet Rural, though not free from symptoms, are rather highly resistant, Among European varieties are some which are symptomless carriers and some which become systemically necrotic when infected. The symptom on "susceptible varieties is mottle, varying from, a mild diffuse chlorosis in some to very pronounced mosaic in others. Warm temperatures promote masking of symptoms; cool weather favours severe mottling and stunting. Affected plants produce undersized tubers, which usually show no visible symptom other then smallness.

- *Rugose Mosaic.* This disease is incited by the veinbanding virus, and in the case of symptomless carriers of the latent-mosaic virus the synergistic effect of the latter is the chief factor in symptom production. Mottle is relatively inconspicuous in most cases. Rugosity and curling of the surface of the leaflet together with some stunting are the most prominent early symptoms. Necrosis usually follows, starting as individual spots on leaflets and progressing until leaflet, and eventually the entire leaf, dies, some types dropping but often clinging to the stem. Necrosis is most severe on the Jowest leaves and progresses upward slowly, a tuft of rugose leaves often persisting at the tip until the top dies prematurely. The tubers ordinarily show no symptoms other than the reduction of size induced by damage to the top.

THE VIRUSES

1. Latent-mosaic virus.
   - Potato mottle and ring-spot virus.
   - Potato-acronecrosis virus. Potato virus X. Potato virus 16.
   - *Solanum* virus 1 R.M. Sm.
   - *Annulus dubius* Holmes.

One of the first evidences of this virus was that noticed hy Johnson when. Juice extracted from apparently healthy potatoes was found to incite a mild mottle of tobacco and tomato and sometimes ring spot on tobacco. Koch later separated a mottle strain and a ring-spot strain. Salaman defined several strains of the virus and was the first to show that one strain, if inoculated in advance in tobacco, protected against another even more severe strain.

Investigators in several countries have described strains, many of which differ in type of symptom incited and in severity of disease development. While cross-protection is rather complete on tobacco, it is not uncommon to find two or more strains in a given potato plant. There are indications that the virus mutates frequently in the potato, and "new" strains are isolated readily when special techniques,are employed.

In general the virus is relatively highly resistant to physical agents. The inactivation points are for dilution, 1-10,000; heating, 66°C.; aging in vitro, 2 to 3 months. The virus was isolated in crystal form by Bawden and Pirie. The host range includes tomato, tobacco, pepper, petunia, henbane, jimson weed, and black nightshade. No insect vector has been found. The virus is transmitted readily by mechanical means.

2. Mild-mosaic virus.
   - Super mild-mosaic virus.
   - Crinkle-mosaic virus.
   - Potato virus A.
   - *Solanum* virus 3 K. M. Sm.
   - *Marmor solani* Holmes.

The virus is- transmitted mechanically and by several species of aphids. The physical properties include thermal inactivation, 50°C.; inactivation by aging in vitro, 24 hr.; inactivation by dilution, 1-100. The host range includes tobacco and several other solanaceous species.

3. Veinbanding virus.
   - Potato virus Y.
   - Solanum virus 2 K. M. Sm.
   - Marmor upsilon Holmes.

This virus is now known to affect a wide host range, including black nightshade,- jimson weed, tomato, henbane, petunia, and tobacco, in which it incites a distinct veinbanding symptom. The physical properties include thermal inactivation, 52°C.; inactivation by aging in vitro, 36 hr.; inactivation by dilution, 1-1,000. It is transmitted mechanically and by several species of aphid.

DISEASE CYCLES

The chief means of perpetuation of the three potato viruses is by way of the seed tuber. Local dissemination of the latentmosaic virus is by mechanical contact, aided in the field by man, animals, and implements, and also occurs in the preparation of seed stock by the cutting knife. Aphid transmission is the major means of spread of mild mosaic and veinbanding viruses during the growing season.

An early observation with mild mosaic was the fact that crops of Bliss Triumph grown in the southern states in midwinter from northern certified seed often had a very high incidence of mosaic although little of the disease was seen in the seed field the previous year. Johnson pointed out that air temperature had an effect upon the expression of symptoms. Tompkins studied the problem further and showed that relatively short daily exposures of infected plants to 23°C. were sufficient to mask the symptoms.

This explained why the disease was often overlooked in seed fields in warm midsummer weather. Latent potato mosaic is also favoured by cool temperature in symptom expression on susceptible varieties. Rugose -mosaic, on the contrary, is enhanced by warm weather and suppressed by cool weather. "Varietal Resistance" Complete immunity from latent mosaic is known only in one seedling. This has not shown any signs of infection by any strain of the virus tested. A considerable number of varieties are top-necrotic, and since infected individuals are - eliminated, the respective variety is immune for practical purposes. Many varieties are symptomless carriers; others are mixed populations of symptomless carriers and susceptibles. In the United States all old varieties are symptomless carriers; most new varieties are mixtures of those in the latter class and susceptibles.

The key to the symptomless-carrier picture in different varieties lies in the fact that all varieties derived from seedlings start out as latent-virus-free stocks and gradually the population becomes infected. Those individuals which become infected by a sufficiently mild strain of the virus become symptomless carriers, and in the course of seed selection they become the bases of clones which are perpetuated. The surviving stocks of old varieties are thus found to be infected generally with a relatively mild strain which is protective against severe strains. In new varieties which are not in the immune or the hypersensitive class, one finds some plants which have become infected by a mild protective strain and other plants which are entirely free from a latent-mosaic virus. When the plants in the latter group become infected by a severe strain, they develop acute symptoms; if they, by chance, become infected by a mild strain, initial symptoms may appear, but the next vegetative progeny of such plants are mostly plants which are symptomless carriers.

According to Stevenson et al. immunity in controlled by two genes, A and B, both of which are necessary for expression of the character. Cadman and Cockerham attribute the top-necrotic reaction to a given strain of latentmosaic virus to the -control of a single gene.

While certain European varieties are known to be top-necrotic when infected by the mild-mosaic virus, most varieties which are designated as resistant are only field-resistant, since they develop typical symptoms if the virus is introduced by grafting. Among the field-resistant varieties are Katahdin, Chippewa, Sebago and many other subsequent introductions by the U.S. Department of Agriculture. Very susceptible varieties are Bliss Triumph, Red Warba, and Green Mountain. While seedlings differ in their reaction to rugose mosaic, none of the commercial varieties in use in the United States is highly resistant.

In the breeding programmes in the United States the immunity of S41956 is being introduced into new varieties. In the British Isles the necroticreaction character with respect to all three viruses is being used in breeding lines.

CONTROL

When varieties susceptible to one or another of the mosaic diseases are grown, the first essential control measure is the use of seed tubers low in virus content. This is made possible through the production of certified seed which has a very low incidence of infected tubers. Old varieties are generally symptomless carriers of the latentosaic virus, and varieties field-resistant to mild mosaic can be had for most needs.

Since there is evidence at hand from experiments in the United States, England, and Australia that yield is reduced somewhat in a symptomless carrier of latent-mosaic virus when compared with virus-free clones of the same variety, attention is being given in Great Britain and in Australia to the selection and maintenance of potato stocks free from latentmosaic virus as well as from other viruses. It remains an open question whether the gain from such a procedure will justify the cost involved.

BEAN MOSAICS

A mosaic disease of bean was first noted in 1899 by Iwanowski in Russia. It is now known that several distinct viruses may affect the common bean. The disease incited by a seed-borne virus and now known as common mosaic received first attention. In 1934 Pierce defined a second virus and named the disease it incited yellow mosaic. The two viruses have been shown and they differ chiefly in host range and in the fact that one is seed-borne and the other is not. They are the most important mosaic diseases on the crop and often cause severe reductions in yield.

They have been found throughout the United States and in many parts of the world. Curly top of bean is very severe in the United States west of the Continental Divide. It is incited by the sugar-beet curly-top virus. Southern bean mosaic and pod mottle have symptoms some of which resemble those of common and yellow mosaic, but the properties of the viruses are quite distinct."Symptoms" On susceptible varieties common mosaic causes various degrees of veinclearing and mottle on the leaves, the greener portions of the pattern often becoming decidedly puckered. In some varieties extreme malformation of leaves may occur.

Pods ordinarily show no signs, but under some conditions chlorotic spots may appear, and sometimes a glossy sheen of immature pods is noticeable. The symptoms of yellow mosaic overlap those of common mosaic, and the two diseases are often difficult to distinguish in the field. In general, yellow mosaic is more severe then common mosaic, and it occurs on many varieties which are resistant to the latter. In plants affected with one strain of the yellowmosaic virus, pods are distorted severely. With each disease, plant growth is reduced; blossoms tend to drop; and pods are shorter then normal.

Under exceptional circumstances the variety Corbett Refugee and varieties derived from it, which are resistant to common mosaic, show extrem enecros is of leaves, stems, and roots due to infection by that virus.

This has been shown to be a reaction which these varieties exhibit when they become systemically infected through the phloem. The symptoms are similar to those of a virus disease described by Jenkins as black root.

THE VIRUSES

Common mosaic: Bean -virus 1.

- *Phaseolus virus* 1 K. M. Sm.
- *Marmorphaseoli* Holmes.
- Yellow mosaic: Bean virus 2.
- *Phaseolus virus* 2 K. M. Sm.

The properties of bean virus 1 have been determined by Fajardo and Pierce. Reddick and Stewart reported seed transmission in 1919. They pointed out in 1918 and 1919 that the variety Robust was highly resistant to mosaic in New York. The resistance of Robust was observed by other workers in other states. Pierce noted a mosaic of Robust in Wisconsin and from infected plants recovered a vir;us-which had similar physical properties but was not seed-borne. It affected some bean varieties and other species which had been shown to be resistant to bean virus 1. He described the new virus as bean virus 2. An important means of differentiation was based on the fact that bean virus 2 was infectious to Robust, Corbett Refugee, and Idaho Refugee and was not seed-transmitted, while with bean virus 1 the opposite was true.

In 1943, Richards and Burkholder described a virus in New York which resembled bean virus 1 in being seed-borne and noninfectious on Corbett Refugee and Idaho Refugee, but it was infectious on Robust and Michelite, a new variety derived from Robust. What appears to be the same virus was described by Dean and Hungerford in Idaho in 1946. This virus as regarded by Grogan and Walker as a strain of typical bean virus 1 and was designated by them as Burkholder's strain. Gtogan and Walker in 1948 described from Wisconsin anew strain of bean virus 2, which differs from the typical strain in that infection results in severe pod distortion on bean and the virus does not infect sweet clover. Zaumeyer and Thomas described a new virus on bean, which they called the greasy-pod, or shiny-pod, virus. Grogan and Walker regarded it as a strain of bean virus 1.

The appearance of various other strains of,bean virus 1 and bean virus 2 is likely. That the two viruses are closely - related was shown by cross-protection tests. The two criteria of distinction between the strains are not on the basis of symptoms, as in tobacco mosaic, but on the basis of host range and seed transmission.

The differential reaction of bean varieties to the strains is of particular importance because the major approach to control of the two diseases has been through breeding for resistance. Both viruses are transmitted by aphids. The symptoms of common mosaic tend to be masked, while their fullest expression takes place between 20° and 28°. Bean virus 1 may be carried with the pollen. The percentage of infected seed varies with the time of -infection of the plant in relation to maturity of a given pod on the plant. Generally speaking, little more than 50 per cent of seeds become infected.

VARIETAL RESISTANCE

Robust was the first- variety which showed high resistance to bean virus. It was developed in Michigan by Spragg from a single-plant selection in Pea bean and came into use commercially about 1916. Robust was outstanding in that plants failed to harbor the virus in any degree when inoculated. Pierce developed in Idaho several strains of Great Northern field bean which were equal to Robust in resistance. Corbett Refugee was derived by Ralph Corbett in Montana from a highly resistant individual in the very susceptible variety. Stringless Green Refugee.

This was used as a breeding parent to develop the resistant varieties Idaho Refugee, Wisconsin Refugee, and U.S. No.5

Refugee. Idaho Refugee and U.S. No.5 Refugee varieties have become widely used as snap beans.

The inheritance of resistance in Robust was studied by McRostie and Parker and that in Corbett Refugee by Parker and Pierce. These workers did not arrive at definite conclusions as to the nature of inheritance. Pierce showed that in the F_2 pregnancy from a cross between Robust and Corbett Refugee, both resistant to bean virus 1, segregates susceptible to this virus appeared. This indicated separate genes for resistance in the two varieties.

The genetics of resistance in these two varieties or in varieties derived from them was cleared up by Ali. He found resistance of Idaho Refugee and of U.S. 5 to be dominant to susceptibility of Stringless Green Refugee, while the resistance of Robust was recessive to the susceptibility of Stringless Green Refugee. Resistance and susceptibility in these three types are dependent upon the interaction of two alleles, *Aa and Ii.*

A is a dominant gene for susceptibility.

I is a dominant inhibitor which prevents the expression of susceptibility due to *A*. *a* is the recessive gene for resistance.

AA II is the genotype of susceptible Stringless Green Refugee.

AA II is the genotype of Idaho Refugee, U.S. 5, and Corbett Refugee, in which phenotypic expression of resistance is due to the dominant inhibitor *I*.

AA II is the genotype of Robust, Michelite, and probably Great Northern) in which phenotypic expression of resistance is duc to the recessive allele.

Grogan and Walker showed that although varieties such as Idaho Refugee and U.S. 5, which derived their resistance from Corbett Refugee, ordinarily did not harbor the virus when inoculated, the virus sometimes became systemic and caused severe top necrosis. While the percentage of plants which responded in this way, following the rubbing inoculation technique, was very low, a large percentage of the plants succumbed when they were inoculated through approach grafts with a susceptible variety. Varieties such as Michelite, derived from Robust, on the other hand, remained perfectly healthy when inoculated by the approach-graft method. It appears, therefore, that in the, Corbett type of resistance, where the dominant gene for susceptibility is present, the effect of the inhibitor gene is complete under ordinary methods of inoculation, but when more continuous exposure to the virus is provided through the approach graft, the extreme necrotic reaction follows when A is present. In Michelite, where A is absent, there appears to be no ingress of virus and no reaction.

The genetic basis for this difference between the two resistant hosts was demonstrated by Ali. When Robust and Idaho Refugee were crossed, he found segregation in the F_2 generation into susceptible and resistant classes, as did Pierce, and the ratio was that expected from a dihybrid, i.e., 13 resistant to 3 -susceptible. When the plants in the resistant class were inoculated with bean virus 1 by the approach-graft method, this portion of the population segregated in the ratio of 9 top-necrotic and 4 nonnecrotic.

Wade and Andrus secured monogenic segregation in the F_2 generation of a cross between U.S. No.5 and Stringless Black Valentine. The latter variety apparently has the genetic composition *AA ii*, in common with Stringless Green Refugee.

Pierce pointed out that many varieties which were susceptible exhibited relatively mild symptoms, although the virus increased in the tissues. He placed these in a class designated as "tolerant." There are many varieties in this class in which resistance is of substantial commercial value. All varieties tested were susceptible to typical yellow mosaic, but, many showed somewhat more, tolerance than others.

Control. The only feasible control for common bean mosaic is the use of resistant or tolerant varieties. Several resistant varieties of the type of the very susceptible semi determinate Stringless Green Refugee variety are now available, e.g., Idaho Refugee, Sensation Refugee, and U.S. No.5. In the class of bush-type snap beans are Logan and Rival. Robust and Michelite are widely ~sed pea bean varieties. At the University of Idaho several resistant strains of the Great Northern field bean type have been introduced. It is to be noted, however, that Robust and Michelite and some of the U. 1. series are not resistant to the seedborne Burkholder strain of bean virus 1.

Yellow mosaic is more difficult to control, but since it is not seed-borne, it is not so widely destructive. Since sweet clover is the commonest source of inoculum, a certain degree of control can be attained in some areas by not planting beans near sweet clover fields and where it - occurs in headlands and fence rows. Gladiolus may- also be a reservoir of this virus.

PEACH YELLOWS AND LITTLE PEACH

Peach yellows and tobacco mosaic were the first virus diseases to be studied critically. The for disease had been known in this country since 1791. The pioneer Work of Erwin F. Smith, done in the late 1880's and early 1890's, was a most painstaking and exhaustive investigation. Smith distinguished two diseases of peach, i.e., yellows and rosette. Another virus disease, little peach, was first noted in Michigan in 1893. Kunkel many years later showed that the viruses of yellows and little peach were closely related, and he concluded that they should be regarded as strains of the same virus.

Both yellows and little peach are confined to North America and are found almost exclusively in northeastern United States and southeastern Canada. The peach-yellows virus also incites diseases in almond, apricot, nectarine, and - plum, and the little-peach virus incites a disease on plum. All species of Prunus tested are susceptible to the yellows virus or are symptomless carriers of it, while some plum varieties are symptomless carriers of the little-peach virus. Until control measures were developed, the two diseases were known to destroy practically entire orchards.

Symptoms. A relatively inconspicuous veinclearing is one of the early symptoms of yellows. The leaves at the tips of infected branches turn over in a sickel-shaped manner. The margins of the leaves may roll inward. Upright-growing shoots bearing small chlorotic leaves appear later, and normally dormant buds tend to grow out. Bud growth in the spring is in advance of that in the healthy tree. Fruits ripen prematurely, and the surfaces are coarsely blotched with red and purple spots which extend into the flesh, resulting in insipid flavour. In succeeding years more diseased branches appear, and the foliage becomes more generally yellow to reddish brown, and leaves are more conspicuously rolled. Trees eventually die within 1 to 5 years after the signs of the disease appear.

Little peach appears first as a distortion and a rugosity of young leaves, which tend to become smooth as they enlarge. They are greener than normal at first, later becoming yellower than healthy leaves. Short upright branches in excessive number develop. The internodes remain short, a general stunting follows. In successive years the number of short twigs increases. Fruits remain small and become misshapen. They tend to ripen somewhat later than healthy ones. Trees die within a shorter period than those affected with yellows.

THE VIRUSES

1. Peach-yellows virus.
   - Peach virus 1.
   - Prun8 virus 1 K. M. Sm.
   - *Chlorogenus persicae var. vulgare* Holmes.
2. Peach little-peach virus.
   - Peach virus 3.
   - Prunus virus lA K. M. Sm.
   - *Chlorogenus persicae var. micropersica* Holmes.

Disease Cycle. Smith was the first to demonstrate that the yellows viruses were transmitted by budding and grafting, but he was unable to transmit them by mechanical means. Kunkel published evidence in 1933 to show that the peach-yellows virus was transmitted to about 10 per cent of the trees exposed to the plum hopper, Macropsis irimaculata Fitch. The insect was found also to transmit the little-peach virus. This hopper does not -move by hopping, and it seldom flies.- It feeds principally on the bark of old woody branches of old trees. Its habitat covers approximately the same range as that of yellows and little peach.

When diseased buds are grafted on healthy trees, the viruses move-into the latter in 8 to 14 days. If infected buds are grafted near the base of the tree, a period of a year or more may elapse before signs of transmission and infection appear, while inoculation in this manner near the top of the tree may result in a much shorter incubation period. Kunkel attributed this to the fact that the viruses move downward almost exclusively. The shortest incubation period for peach yellows was 78 days.

He also established the fact that the insect did not become capable of inoculating the virus until some days after it had fed upon a diseased plant. Hartzell found the range of this nontransmissible period to be from 7 to 26 days. Kunkel showed that plants infected with the yellows virus were immune to the little-peach virus and vice versa. On the basis of these facts he regarded them as strains of the same virus. The peach-rosette virus had no such relation to either the,yellows or the little-peach virus.

Trees infected with the yellows or the little-peach virus were cured by inserting them in a tank of water at 50°C. for about 10 min or by keeping them in an air temperature of 34.4 to 36.3° for about 25 days, the time required for a complete cure in the latter treatment increasing with the size of the tree.

CONTROL

Eradication of infected trees and disposal by burning as soon as they appear are the most effective means of control. Healthy trees may be set in place of removed ones, since the virus is not soil-borne. Rigid nursery inspection and eradication in the orchard have been successful in keeping yellows and little peach from becoming major diseases. Manns recommends removal of all wild plums within 1 mile of the orchard.

ASTER YELLOWS

One of the most widespread virus diseases in the yellows group is that known as aster yellows. It acquired this name because early in this century it became consficuous and destructive on China aster in many parts of the United States. The disease now is known in many other countries. One of the first descriptions on the aster was by R. E. Smith in 1902 in Massachusetts. An insect vector was first reported in 1924 by Kunkel, who transmitted the virus to many species over a range of some 150 genera in 38 families. Sever in and associates extended the range still further. Aside from aster and many other ornamental flowering plants, common economic hosts are carrot, celery, lettuce, endive, parsley, salsify, onion, and potato.

SYMPTOMS

The first sign of the disease on aster is veinclearing in the young leaves. Leaves become more chlorotic as they develop. The chlorosis may affect leaves unilaterally, and sectors of an entire plant may show symptoms while the remainder appears healthy. The normal colour of the petals is changed, and a greenish tinge commonly appears instead. "Breaking" of flower petals has been observed in annual phlox. Aster leaves which are mature when symptoms appear on the plant do not change in colour. At the axils of such leaves and of those which show symptoms, dormant buds are stimulated, and yellowish thin branches appear with internedes longer than usual. This type of proliferation is one of the striking symptoms of the disease. The main stem of the plant is stunted rather than stimulated. Leaves are, more or less deformed as they grow. Flowers are misshapen and usually worthless. Necrosis of the stem a short distance below the apical bud may, appear in the later stages of the disease.

The chlorosis, proliferation, malformation, and stunting occur in some degree in other plants which are affected. In lettuce the yellowing and distortion of leaves led to the name "rabbit's ear long before the causal nature of the disease was known. The proliferating phase of the disease promotes bolting, and in this way it becomes a very destructive disease on this crop. In carrots excessive proliferation of axillary buds at the crown leads to bushy yellowish tops. The quality of the root is reduced, and secondary rootlets are excessive in number and are malformed. The crowns of diseased plants are predisposed to soft-rot bacteria in moist weather. When affected carrot roots are use for seed propagation, flower umbels are distorted, and the plants often die before seed is mature. The chief damage to onion is in the seed-plant stage, in which malformation and sterility of flowers occur.

The purple-top disease of potato is due to infection by the aster-yellows virus, Young upper leaves become upright, and cessation of apical growth of the stem is noticeable. Leaflets do not expand normally; they roll upward along the edges and become chlorotic.

In varieties which normally contain purple pigment in stem and leaf, the leaflets take on an abnormally purple colour along the edges. But proliferation and axillary tuber formation at leafaxils may follow. Many affected plants develop a permanent wilt. Necrosis occurs in the vascular system in the lower part of the stems of wilted plants and in the vascular and cortical tissues of the stem ends of tubers. In hills of affected plants, tubers are inclined to become flabby and sometimes produce very thin, weak sprouts when they germinate. The Causal Virus

- Aster-yellows virus.
- *Callistephus* virus 1 'K. M. Sm.
- *Chlorogenus callistephi* Holmes.

Two strains of the virus are known to occur, the chief difference being in their infectivity on celery. The strain studied by Kunkel did not affect celery; that occurring in western states and studied by Severing was infectious to that host. The strains are commonly referred to as the eastern, or New York, strain and the western, or California, strain, respectively.

DISEASE CYCLE

The first insect vector to be reported was the six-spotted leaf hopper, Macrosteles divisus Uhl. More than a dozen other species of leaf hopper are now known to be vectors of the western strain.

Eunkel noted that both nymph and adult leaf hoppers are unable to transmit the virus immediately after feeding upon infected plants. Although 10 days must elapse before the insects become inoculative, they are known to remain viruliferous for 100 days or longer. The virus is not transmitted mechanically from plant to plant, but Black succeeded in transmitting it mechanically from insect to insect. A lapse of 11 to 45 days was required before, insects to which the virus. Vas transferred mechanically were effective in inoculating plants. Evidence was secured that a hundredfold increase of the virus had taken place in the body of the insect 6 days before the latter was able to transmit the virus to aster plants.

When colonies of infective hoppers were subjected to heat treatments of 1 day or longer at 31 to 32°C., many lost their ability to transmit the virus permanently, but some regained that ability. When they are exposed for 12 days or longer, all lost permanently the pir ability to transmit the virus. When poppers recovered transporting ability, the virus was sometimes altered so as to incite only mild symptoms on the host, a character which was retained,by the virus in successive plant passages. It was suggested, that midseason decline in transmission of the virus followed by a late season rise might be due to the reduction in virus in viruliferous insects during midsummer heat. This has not been demonstrated in field experiments.

When infected periwinkle plants were kept at temperatures of 38 to 42° for 2 weeks, or when they were immersed in a water bath at 40 to 45° for a few hours, the virus was completely and permanently inactivated. Such cured plants could be reinfected in the usual way, however. Infected plants of Nicotiana rustica L. were cured when kept at 40° for 3 weeks.

It is a common experience of investigators that, although the virus is transmitted to a given host species and typical symptoms follow, it is not recovered when non-infective hoppers are fed on such plants and transferred to susceptible host plants. Severin and Haasis reported disease symptoms in about 50 per cent of the potato plants inoculated, but they failed to recover the virus from any of the plants. Other investigators have had similar experience with this host. Younkin and Self and Darling, however, recovered the aster-yellows virus from purple-top potato plants by grafting.

The virus ordinarily overwinters in northern states in perennial hosts. The time of appearance and the extent of the disease are dependent upon the insect vectors. In western New York M. divisus commonly hibernates on winter grain, probably in the egg stage, and possibly occasionally in the adult stage. The incubation period in the host plant is from 10 to 40 days or longer. It is not Uncommon for the disease to be abundant in one host plant, while another potential host nearby is practically free from disease. This may be explained in part by the preference of the vector for one host as compared with another. The incidence of disease is not always correlated with the population of the vector. The relative abun., dance of virus reservoirs and the conditions which attract or fail to attract insects to those reservoirs are important factors in determining disease incidence..

In a study of the disease on lettuce and endive on Staten Island, New York, Linn found weed barriers to be the chief reservoirs of the virus. Wind direction influenced the spread of the vector, but ordinarily the in sects did not migrate much more than 200 ft. in 4 weeks.

CONTROL

The disease is so widespread and severe on China aster in the upper Mississippi valley that commercial and home-garden culture of this plant has been largely abandoned. Commercial florists control the disease successfully by growing asters under frames completely "covered with shade cloth. When DDT is applied to outdoor crops such as carrot at free quent enough intervals and beginning early enough in the season to reduce hopper populations to a minimum, partial control has been accomplished. Linn found that on Staten Island control in lettuce and endive was commercially practical by screening cold frames, transplanting to areas separated 200 ft. or more from weed borders, or destroying such weeds with sodium chlorate crystals, and dusting the transplanted beds weekly with pyrethrum-sulfur or 1 per cent rotenone sulfur. Ashdown and Watkins secured reduction in the aster-yellows disease in lettuce by dusting with DDT.

In production of carrot seed in regions where aster yellows prevails, it is essential to produce the roots in locations where weed hosts do not occur or where they can be destroyed and the growing crop can be protected from insect vectors.

CURLY TOP OF SUGAR BEETS

A serious disease of sugar beet known as western blight was recognized in California before 1900, and in that year it was found to be present in all western states where beets were grown. In 1905 the disease was associated with the presence of the beet leaf hopper and for a time was considered to be the direct effect of attack by that insect. In 1915 Boncquet and Hartung reported that when leaf hoppers taken from the wild and not previously in contact with beets were allowed to feed on beets, no disease resulted. When they were next transferred to curlytop beets for from 3 to 7 days and removed thence to healthy beets, the latter contracted typical symptoms. This was the first sound evidence that the disease was due to a transmissible entity and not to the direct effect of the insect.

The curly-top disease occurs throughout the beet-growing areas west of the Continental Divide and occasionally, but not severely, in regions immediately east of the Divide. It is confined to the North American continent. A similar but distinct disease occurs on sugar beet in Argentina. For the first three decades in the century curly top increased-in destructiveness until the sugar-beet industry was seriously crippled. The development of resistant varieties initiated by Carsner about 1924 has continued up to the present. Beginning about 1933 with the introduction of variety U.S. 1, resistant varieties have been introduced froip time to time and have reduced greatly the damage by curly top.

The virus has many wild and economic hosts. The former serve to provide a season-to-season reservoir. Among the economic hosts in which curly top is still a major disease are table beet, chard, spinach, tomato, bean, muskmelon, squash, and pumpkin. Sever-in found 75 species included in 48,genera and 18 families to be naturally infected by the curly-top virus in California.; Freitag and Severin transmitted the virus experimentally to 92 species of ornamental flowering plants in 73 genera belonging-to 33 families.

SYMPTOMS

If sugar beet plants are infected when quite young, they die rather promptly. When half-grown plants are infected, they show throughout the remainder of the season the range of symptoms by which the disease is known. The leaves which are mature when the plant becomes infected show no pathological symptoms other than to turn yellow and die eventually.

The younger leaves roll inward, while, the laminae tend to pucker in blister-like elevations. Veinclearing appears in the youngest leaves, and this is followed by swelling on the veins on the lower side of the leaves, which assume the appearance of knot-like galls or nipple-like papillae. An exudate of liquid appears on petiole, midrib, and the lower sides of the veins. This is clear and viscid at first, becoming dark and sticky later, often drying to form a brown crust. The affected leaves may remain dull dark green, but they eventually turn yellow, then brown, and die prematurely. The environment influences the type and rate of symptom development.

Necrosis occurs in the phloem of leaf, petiole, and root, and in advanced stages of the disease dark concentric rings are visible in the flesh of the sugar beet when it is cross sectioned.

In tomato, leaflets of infected plants roll upward along the midrib and become thickened and crisp as the leaf curves downward. The leaf tissue gradually turns yellow, while the veins take on a purple tinge. The plant assumes an erect habit and remains stunted. In squash young leaves have a savoyed surface or an upward roll; internodes remain short; many blossoms fail to set; and the tips of runners turn upward. Plants infected when very young are killed rapidly without the above symptoms in warm weather. Older vines decline gradually and die prematurely. In bean, young affected plants may show killing of the growing point and yellowing of the primary leaf, followed by death of the plant. When older plants are infected, there is a downward marginal rolling of trifoliate leaflets, followed by brittlenees, dwarfing, and puckering. The whole plant may become stunted and unproductive.

THE VIRUS

- Beet-curly-top virus.
- Sugar-beet virus 1.
- Beta virus 1 K. M. Sm.
- *Chlorogenus eutetticola* Holmes.
- *Ruga verrucosans* Carsner and Bennett.

The physical properties of the virus as determined by Severin and. Freitag and Bennett are inactivation in vitro in extracted juice, more than 7 days; by dilution, above 1-20,000; by heating, 75 to 80°C. "The virus is unusually resistant to alcohol and acetone and survives in solutions up to pH 9.1. In the phloem exudate, it withstands freezing up to 18 months. It passes _commonly used porcelain filters. It remained active in thoroughly dried young sugar-beet plants for 8 years, leaf hoppers for 6 months, and in dried phloem exudate for to months.

DISEASE CYCLE

The virus is perpetuated -between crop seasons in wild hosts and it may overwinter in the leaf hopper. In semiarid regions such as southern Idaho, abandonment of land and overgrazing often tend to encourage plants which are good overwintering hosts for the insect and which may serves as reservoirs for the virus as well. The injury to beets by transmission of the virus by the insect is dependent primarily upon the time and magnitude of spring leaf hop per movement-into cultivated fields. Factors which determine the magnitude are size of fall migration to winter feeding areas, winter survival, and spring reproduction; these in turn are influenced by the type and amount of fall, winter, and spring host plants. The date of the spring migration to beet fields is influenced by the prevailing temperature. The direction of the migration is determined by prevailing winds, which are usually from the west.

The only vector known is the beet leaf hopper; Eutettix tenellus Baker. The extract from diseased plants or infective hoppers may be used as a source of inoculum, experimentally, by allowing hoppers to feed upon it and by transferring them to the host plant. The virus has not been transmitted mechanically. Early investigators found that hoppers were not infective until 1 to 44 days after feeding on diseased plants. Severin secured transmission in a small percentage of cases within 20 min. Bennetit and Wallace were not successful in securing transmission in less than 4 hr., although the insect picked up the virus within 1 min. Freitag and Bennett and Wallace concluded that the virus does not multiply within the insect.

Carsner secured an attenuation of the virus by passage through *Chenopodium murale* L. Lackey showed that virulence could be restored by passing the attenuated virus through Stellaria media Cyr. Giddings isolated 10 distinct strains of the virus. He also presented evidence to show that infection was dependent upon mass action of the virus introduced.

The relation of the virus to the host tissue and the reaction of the latter to infection has been studied critically by Esau. The virus in leaves apparently is contained entirely in the phloem during the time it increases and is transported. After one or more of the primary sieve tubes have differentiated, the first histological signs appear in the form of hypertrophy and death of the pedderm and phloem cells adjacent to the sieve tubes. In cells somewhat farther removed from the sieve tubes, hyperplasia occurs; the cells involved go through some of the changes characteristic of differentiation, of sieve tubes, but normal maturation is not complete.

These primary phases of hypertrophy, necrosis, and hyperplasia are followed by secondary necrosis in the abnormal sieve tubes aped their companion cells. The parenchyma cells around the centres of secondary necrosis next undergo secondary hypertrophy and hyperplasia, resulting in preparation resembling wound periderm reactions. Hypertrophy and hyperplasia are not Gonfined to the phloem. The mesophyll cells adjacent to the phloem are stimulated to grow and divide so that they become closely packed together without intercellular spaces, while the chloroplasts show signs of degeneration. This internal change in the tissue is responsible for the macroscopical veinclearing noted as an early symptom. Further hypertrophy and hyperplasia in cortical tissue result unequal information in leaf and petiole tissue, including the formation of the characteristic protuberance. Lackey demonstrated the virus, in the ineristematic tissue of the root tips.

Bennett showed that if the virus is introduced into the distal end of a beet leaf in the dark, it often fails to move out of the leaf in a period of several days, whereas when it is inoculated in a green leaf in the light, it moves out in as short a period as 4 hr. He believed that the movement of the virus has little relation to its multiplication or to its concentration but is correlated with food transport and takes place entirely in the phloem.

VARIETAL RESISTANCE

As already indicated, numerous varieties of sugar beet resistant in various degrees to curly top have been developed and released for use in various areas to which they are adapted. All such varieties are carriers of the virus and may show signs of the disease under optimum conditions.. They are genuinely tolerant, however, and the amount of phloem necrosis is greatly reduced, as shown by Bennett and Esau, while the actual amount of virus present in the tissue is less, as shown by Giddings. Lackey found that in a fairly resistant Great Northern bean the virus was promptly inactivated in the root tip, while it persisted in the susceptible Bountiful variety. In heritance of resistance in bean has been studied by Schultz and Dean.

CONTROL

Band and crop management and controlled grazing are important measures in reducing as much as possible the magnitude of the spring migration of leaf hopper from overwintering areas to cultivated crops. Early planting and liberal fertilization of resistant varieties of sugar beet are desirable. At present certain-susceptible varieties and species must be avoided in areas where the chances of heavy infestation by viruliferous hoppers are great.

Beginning with the resistant variety of sugar beet U.S. 1, there has been developed a succession of varieties of greater agronomic value and adaptability to various areas. In southern areas, where culture comes during winter months, easy-bolting varieties must be avoided, although such a variety may be quite satisfactory for a summer crop northern areas. The variety U.S.,22 occupied most of the Idaho acreage in 1942, and an improvement of this line was released in 1946 as U.S. 35. In the same year a new non-bolting curly-top- and downy-mildew-resistant variety U.S. 36, was released-Jar use in the San Joaquin valley and in the Imperial valley of California.

Dana apointed out that Umatella Marblehead and Yakima Marble head varieties of squash are spaturally highly olerant to curly top and are of particular value in the Pacific Northwest.

In southern Idaho the Common Red Mexican bean is resistant to curly top, while the more widely grown strains of the variety Great Northern are susceptible to curly top but resistant to common mosaic. From hybrids between these two parents a new variety of the Great Northern type resistant to both virus diseases, has been introduced as Great Northen U.I. 15. Pioneer is a curly-top-resistant snap bean introduced by the U.S. Department of Agriculture and the Oregon Experiment Station.

SPOTTED WILT

The disease known as spotted wilt was first defined as such in 1919 on tomato in South Australia, where by 1925 it had become a serious disease causing enormous losses. It has since been found oil many crop plants and wild species in continental United States, Hawaii, Great Britain, the European continent, South Africa, Brazil, Argentina, and New Zealand. Among the economic host plants besides tomato are tobacco, potato, pepper, lettuce, endive, celery, spinach, pea, and numerous ornamental species including sweet pea like: zinnia, nasturtium, poppy, dahlia, calla lily, orchid, chrysanthemum, petunia, aster, schizanthus, salpiglossis, cineraria, and gloxinia. The yellow spot of pineapple destructive in, Hawaii was shown to be incited by the spotted-wilt virus. A list of hosts was compiled by Smith

The virus concerned is distinctive in that it is transmitted by species of thrips, rather than by one of the more common groups of vectors of plant infecting viruses.

SYMPTOMS

Since the disease is most widely known and commonly destructive on tomato, the symptoms on that host have been studied most extensively. In all hosts, it is well to not the symptoms may be mild under certain conditions and quite different, symptomatically and more severe in another environment. On tomato one of the first characteristic signs is bronze-coloured marking on the upper side of young leaflets, which may be accompanied by a slight downward curling of the leaves.

The intensity of the bronzing varies, and it is undoubtedly influenced by the environment; it may extend to petiole, stem, pedicel, and calyx. Growth may be checked permanently or only temporarily. Upward marginal rolling and stiffening of leaflets follow, and small circular necrotic spots commonly appear on the leaves under optimum conaltions. Necrosis extends to the stem near the growing tip, causing the latter to wilt and die Some mottle, chlorotic spotting, and malformation may occur on leaves.

On the green fruits, yellowish spots up to 1 em. in. diametre appear, often with distinct concentric zones 01 shades of yellow or bronze alternating with green and later with pink or red. The zoned fruit spots are probably the most characteristic sign of spotted wilt on tomato.

On tobacco circular necrotic spots of various sizes appear on young leaves. A pronounced mottle may follow, often confined largely to the lower half of the older leaves, while necrosis of the leaf veins adds to the irregularity of the mottle pattern. Plants are stunted, and leaves may droop gradually and die. Leaves which are mechanically inocuiated show local lesions up to 4 or 5 mill. in diametre.

Symptoms vary on the numerous host plants, and often times symptoms are so variable as to be -of little diagnostic value. Various types of zoned chlorotic and necrotic spots run through the symptoms syndrome.

THE VIRUS

- Tomato-spotted-wilt virus.
- *Lycopersicum virus* 3 K. M. Sm.
- *Lethum australiense* Holmes.

The virus is inactivated by heating at 42 to 45°C. for 10 min.; at dilutions of 1-100,000; by aging in crude plant sap at 20°C. for 5 hr. Longevity in vitro is increased by low temperature and by addition of reducing agents to the extract, since the rapid inactivation of the virus in extracted sap is correlated with the activity of dehydrogenases in the mixture Insect vectors include onion thrips, black carnation thrips, *F. occidentalis* Perg., and *F. moultoni* Hood.

DISEASE CYCLE

The virus is perpetuated in plant hosts. In areas with severe winters reservoirs consist of plants growing in greenhouses or of outdoor perennial hosts, of which dahlia is an important one. In areas with mild winters, winter annuals are common reservoirs. With a wide host range among ornamental plants, dooryard flowers are often the source of inoculum by which commercial crops are infected. The disease is severe on tomato and other vegetable crops during the cool season in the San Francisco Bay region of California.

Although the virus may be transmitted mechanically, insect vectors are the most important means of dissemination. Only the larval stage is capable of picking-up the virus from infected plants. Insects do not become infective until 5 to 9 days after feeding, and they remain infective into the adult stage. Sunlight exerts an inhibitory effect upon the development of spotted wilt lesions on tobacco.

CONTROL

In greenhouse culture the infected—plants which, though mildly affected, serve as a source of inoculum should be segregated and the control practiced.

In outdoor culture in areas where spotted wilt occurs regularly in severe form, some reduction of disease is obtained by elimination of overwintering hosts, placement of fields some distance from potential reservoirs, particularly dooryards, and control of thrips by the most effective practical method.

Some lines of red-currant tomato are highly resistant to spotted wilt and have been used as parents in breeding programmes in California and Hawaii, where the disease is often a limiting factor in -tomato production. The Pearl Harbor variety introduced by the Hawaii mxpelliment Station is highly resistant to spotted wilt. Holmes found that the resistance of Pearl Harbor did not hold up in New Jersey, while a monofactorial resistance found in an Argentine variety was effective in that area.

6

Anatomy of Diseased Plants

REGENERATION OF TISSUES

In many cases the tissues of growing parts of plants, destroyed by traumaticaction, can be regenerated from tissues of different origin. The conventionally distinct derivatives of the dermatogen can be replaced from those of the periblem and those of both the dermatogen and the periblem can be reconstituted from those of the plerome.

Exposed young cortex cells may directly become epidermal by thickening their outer wall, acquiring a cuticle and even forming hairs or stomata. This occurs quite normally in citrus leaves, in which during development the upper epidermis splits and is replaced from the mesophyll, sometimes after a few cell divisions. Still more remarkable is the replacement that has been effected under suitable conditions of the entire lost halves of a young tuber, stem or petiole, in form and structure very like the normal.

A perforating wound involving the central cylinder may become surrounded, not only by a cork layer, but by a deeply seated cambium which may cut off xylem on the side away from the wound and phloem towards it, the part between phloem and cork coming to resemble a cortex. In perforating needle wounds vertically down to the pith in young sunflower heads the cavity can become lined with a complete vascular ring, and bundles of the hard bast type develop on the side of the phloem towards the cavity; then follows a thin-walled parenchyma resembling a cortex, and finally, lining the cavity, an epidermis forms, sometimes with hairs like those on the surface of the plant. Even natural hollows in the stem of Brassica oleracea can develop a 'cambium with phloem next the cavity and xylem outside, and the injection of certain chemicals has caused the same condition in the hollows of young internodes of *Ricinus*. The restoration of lost parts, however, is ordinarily preceded by the development of repair tissue on the denuded surface.

WOUND REPAIR

The commonest precursor of the replacement of lost tissues under moist conditions, especially in herbaceous plants, is callus formation. Callus can be formed in suitable circumstances from any living cell or tissue, and the subsequent changes that it undergoes are independent of its origin.

The living cells bounding the wound below those that have been killed in wounding grow out, multiply, and form a tissue of dividing thin-walled parenchyma. Sooner or later differentiation occurs in this tissue. Its outer margin may become corky by the development of a cork cambium or may consist of a loosely united mass of cells, resembling the so-called hyperhydric tissue. This may arise from a meristem which cuts off roundish, loosely attached cells outside and thin-walled callus-like cells inside, or it may grow without the intervention of a meristem, since callus cells far removed from the level of the wound may continue to divide and form ever-increasing masses of thin-walled parenchyma.

In the deeper part of the callus the first differentiation often is the direct transformation of a callus cell, derived from any tissue, by reticulate thickening and lignification of its walls, into a tracheid-like element. This is one of the key processes in wound repair.

Several of these tracheids often develop close together, forming a little island of woody tissue. Neighbouring groups may unite to form irregular strands; or an island may become wholly or in part surrounded by a meristem which forms tracheids and fibres on its inner side. At a later stage a procambial strand, forming wood and phloem, may develop in the cells bordering the tracheid groups or strands, and may join up with vascular outgrowths from the original conducting system of the organ. A similar process, without the formation of callus, can restore a broken conducting tract - say, in a leaf - cells of the mesophyll in which become changed into short tracheids which can bridge the gap and which, later, can become bordered by a procambial strand forming phloem and xylem.

Very often a large transverse cambium develops below the free surface of a callus and cuts off wood on its inner side and, less frequently, phloem and secondary cortex-like tissue on its outer side. Occasionally a layer resembling an epidermis clothes the surface, especially when a petiole or stem is split longitudinally. In such a case a central cylinder may be reconstituted by a cambium developing in 'the callus and joining up with that of the uninjured half; this forms not only phloem but a more or less normal cortex and epidermis outside.

Finally a callus may form shoots exogenously from a mother cell -or cell group at or near the surface of the callus. This may happen even in a callus derived from the pith. Shoot formation seems to be stimulated by excessive or unbalanced accumulation of food or of growth-promoting substances, for in uncongenial grafts the upper side, where the imperfect union holds back the passage of material from the scion, may develop large callus-like masses, rich in shoots. In the decapitated swollen hypocotyl tuber of cyclamen there is a formation of adventitious cotyledonary leaves. The vascular system which supplies these arises from the division of one or more of the normal cortical cells, and this process extends back to a bundle.

There may be a central mass of xylem surrounded by phloem in these adventitious leaf traces, or xylem and phloem may pursue a wavy course, not necessarily close together, towards the vascular ring" In callus buds the vascular connection may be established in a similar manner. Roots are very often formed in callus but, unlike shoots, they always come from the deeper layers, usually from the neighbourhood of the pericycle but also sometimes from the medullary rays and the external region of the pith. As the horticulturist well knows, different species vary greatly in their readiness to develop roots from the callus of a cutting. Good callus formation does not necessarily mean good root production. It would take us too far to follow the extensive studies that have been carried out during recent years of the part played by phytohormones in the regeneration of roots and shoots and of the practical application of these studies in horticulture.

A second type of wound healing is common and is, indeed, the normal process in woody tissues exposed to ordinary air. In this the repair resembles the normal growth in thickness in that it is the result of the division of cambial cells, the products of which, as usual, soon lose the aptitude to divide, becoming differentiated. The tissues thus formed are sometimes termed wound wood and wound cortex.

If the wound is an open one extending beyond the cambium, or is the result of an amputation as in pruning twigs or lopping branches, the first stage in the process of healing is the blocking of the tissues by the deposition of suberin in the cells, intercellular spaces and cell walls of parenchymatous tissues and an accumulation of an insoluble gummy substance in the cavities of the conducting elements of the wood. The position of maximum formation of this blocking layer is marked by a brown or dark-green band situated at a variable distance below the surface of the wound, near the surface in the cortex, deeper in the pith.

Its rate of development also varies with the stage of activity of the tissues: when growth is active as in the spring and summer it appears in ten days to a few weeks, but in wounds made in the late autumn or winter the completion of blocking is delayed until the spring; young woody tissues are also more rapidly blocked than old. The chemical nature of the blocking substance is not clear. In large parenchymatous masses of tissues it is a product of fatty acids but in woody tissues it is a viscous or resinous substance, usually termed' wound gum' though it is not a true gum. Sometimes a 'gum zone 'forms in the pith and wood of an amputating wound and a band of cork at the same level in the cortex. When wound gum forms in ray or wood parenchyma cells containing starch, the starch first disappears, and it has been suggested that wound gum may be formed from products derived from the starch together with resinous and ligneous substances and pectin. When micro-organisms invade the cut tissues they may stimulate the formation of wound gum.

Gum barriers, whether formed normally as a result of wounding or as a reaction of the living cells of woody tissues to invasion by a parasite, are considered to play an important part in checking the invasion of parasitic fungi through wounds. The readiness with which the form differs in different trees. In the red gum they develop rapidly under fire scars' in several of the layers of the outer sapwood, and resist the attack of Polystictus pergamenus and Polyporus gloves when the unprotected sapwood is heavily decayed. In oak, ash, and Celtis similar fire scars result in early infection by various fungi.

The formation of the blocking layer is followed by an overgrowth of callus mainly derived from the exposed cambium but by no means entirely so. The relative parts played by the cambium and by the other living tissues in reclothing the exposed surface of a stem wound depend to a great extent upon the nature of the wound. When this is a tangential one caused by removing a slice or layer as, for instance, when a portion of the bark is peeled off a stem or a shaving cut out, perhaps including some of the outer wood, there is extensis destruction of the cambium, and regeneration of the bark is affected mainly by outgrowths of callus cells from such exposed living parenchymatous cells as those of the medullary rays and wood parenchyma; only at a later period does cambial activity playa part.

The process has been carefully studied in Malaya in its bearing on the' stripping' method of treating brown bast disease of the rubber tree, where large areas of the affected and surrounding bark are peeled off and the new bark may develop free from the disease. In *Hibiscus* the peeled stem surface begins to show outgrowths of callus cells of relatively enormous size in about three days, none coming at this stage from the cambium which is largely destroyed by the stripping. At the central part, medullary ray callus plays the largest part; at the margins, cells of all the tissues external to the cambium, rays, phloem parenchyma and cortex, are involved. The callus cells divide and come into contact with one another laterally, so that a continuous clothing of callus forms over the wound. In about another three days, under Malayan conditions, the phellogen from the uninjured bark extends inwards into the cortical callus, and phellogen forms in a continuous layer across the mass of basal callus which clothes the major part of the stripped surface.

Beneath the cork thus formed the callus cells are preserved from desiccation, but for a time there is no further differentiation of tissues in them. Then, at the level of the severed marginal cambium surrounding the wound, a level which is now, owing to growth in thickness of the uninjured wood, somewhat above the bottom of the callus clothing the stripped surface, a new cambium forms in the callus. This comes mainly from an extension of cambial divisions inward from the regenerated cells of the cut margin of the original cambial ring involving the callus cells next in contact with these.

The process extends right across the wound, so that the continuity of the cambial cylinder surrounding the stem is fully restored. Thenceforward, the wound covering follows the normal course of growth in thickness of a stem. In this type of repair the wound wood and wound cortex closely resemble the normal tissues.

In an amputating wound, however, especially in older woody stems, there is less scope for the activity of callus-forming cells in the wood, and the bulk of the closing tissue comes from the cut ends of the original cambium and the tissues in its neighbourhood. An annular cushion is formed which gradually spreads over the exposed wood, often leaving pockets of crossed tissue buried at the level of the wound..In this case cambium formation is less regular than in the last, and isolated tracheid groups may form, each bordered by its own separately developed cambium, cutting off wound wood. The most noticeable masses of wound wood arise when the original cambium, without being exposed, is stimulated into activity by injury to some of its cells or to the bordering tissues, from frost or the attack of parasites. Cankers are largely due to an excessive development of wound wood and are a familiar result of frost and of certain types of insect and fungal injury. The pressure from below caused by this deep-seated growth ruptures the cortex, leaving exposed the newly formed wood with sometimes a dead patch at the base.

Wound wood may also develop in callus tissue, from cambiums formed, as already mentioned, in any part of the mass. These may cut off only wood, without phloem. When the meristem develops, as is not uncommon, around a tracheid group or a mass of sclerenchyma or a localised injury, the xylem may be produced either on its outer or its inner side. In the former type there may be a small phloem group in the centre of the nodule, in the latter, either phloem or parenchyma develops outside the cambium. Structures of the same kind are also found in wound bast and even occasionally around the tracheid groups that form in leaf calluses.

In woody plants they may produce knobby isolated masses of wound wood, such as are found in the secondary cortex of the apple and other trees; while the woody nodules, which are the most serious concomitant of the brown bast physiological disorder of rubber in Eastern plantations, are composed of xylem developed on the inner side of a cambium surrounding a little island of uncrossed tissue in the phloem. Very often the wound wood is developed in vortex -like masses of 'fibres and vessels, probably as a result of pressure from unevenly developing meristems or uneven growth at the junction of normal and wound wood. Woody galls often show beautifully grained examples of whorls of this nature.

Wound wood differs histologically from normal wood in having less regular rows, composed of shorter and generally thinner-walled elements, in the usually greater proportion of wood parenchyma and fewer fibres, and in wider medullary rays. Its less solid construction renders it more liable to damage; thus the new wood in frost cankers as readily again injured -by frost, and that resulting from woolly aphis in apples and *Helopeltis* in tea is very prone to the attacks of fungi.

Instead of callus or wound wood forming, a wound may be cicatrized by wound cork. This commonly occurs in all organs of plants in dry air. The cork tends to form parallel to the surface of the wound and often differs from normal cork in having larger cells and thinner walls. When a leaf or a potato tuber is cut, the living cells nearest the wound may enlarge, sometimes with a few divisions parallel to the surface, and their walls become metacutinised or impregnated with a deposit of suberin or of a lignin-like substance or of wound gum.

Below this' blocking layer' the cells elongate and divide transversely, the middle divisions corresponding to a phellogen which cuts off cork cells in rows to the outside and phelloderm rows below. All this part is without intercellular spaces and, in the leaf, the distinction between palisade and spongy parenchyma is lost.

Every living tissue from epidermis to cambium or pith may join in forming the cork layer, which always tends to unite at its margin with the surface covering of the sound part, epidermis or normal cork as the case may be. When the epidermis takes part in the process it divides into cells resembling those of the other tissues of reaction, while collenchyma and even young sclerenchyma cells lose their thickening and react like the rest.

In superficial wounds in roots and submerged aquatic stems, apparently correlated with the presence of a well-developed endodermis, there may be no other reaction than the formation of the blocking layer, without any division of the cells or development of meristems. The endoderm is believed to act as a barrier to the passage from the central cylinder of the solutes that are necessary for meristem formation.

These various processes of healing are not always sharply divided, and more than one of them may sometimes be seen in different parts of the same wound.

ABNORMAL HISTOLOGICAL CHANGES FROM NON-TRAUMATIC CAUSES

Besides the processes that result from traumatic action there is another group of modifications in cell or tissue growth arising from inner conditions in which no gross injury is involved. Only a few examples need be given.

Changes, which may sometimes be deep-seated, can be induced in the cells and tissues of plants merely by the action of environmental conditions, such as moisture or light, or by nutritional factors. It is convenient to distinguish the main basic modifications that may result by the terms *hypoplasy* or under-development, *hypertrophy* or swelling without cell division, and *hyperplasy*, an actual increase in the number of cells. When hypoplasy is total and the organ or tissue does not develop at all, the term *atrophy* is used. These are not the only modifications that can be caused by physiological factors, for the cell membranes and structures such as stomata also respond in varying degree, as for instance in etiolated shoots, sun and shade or submerged and aerial leaves, and over- or under-nourished plants.

Excessive moisture, especially if transpiration is checked, often causes on green stems or leaves the formation of 'intumescences', such as are commonly seen on tomato and other plants in glasshouses. These are usually white protrusions formed of rows of radially elongated, thin-walled parenchyma derived from the epidermis, and mesophyll or primary cortex. Collenchyma can participate in the process, the cells elongating and forming thin dividing membranes.

Large hyperplastic modifications have been induced in some of the biennial *Brassicae* with swollen stems, such as kohlrabi, by preventing flowering. Both cortex and conducting tissue became much enlarged, the xylem of the main leaf trace especially so; this was accompanied by an increase in, and broadening of, the medullary rays, so that a number of narrow bundles were formed. Each of these bundles tended to become surrounded by a cambium developed in the ray cells and deeper part of the wood parenchyma. This cambium formed phloem on the ray side and xylem towards the bundle. Small inverse bundles also formed in the enlarged rays or in the protoxylem region; in these bundles there was a nest of phloem surrounded by a cambium producing a little xylem on the outside. Concentric inverse bundles have also been found to develop in the thickened roots and leaf base cushions of similarly treated savoy cabbage, being in this case situated in the collenchymatous tissue of the inner xylem. In the abnormal tissues that have been thus induced to form, there is a noticeable scarcity of fibres and, where the thickening is due largely to cambial activity, as in swollen side shoots of kohlrabi sometimes, the new tissue on the xylem side consists of thin-walled, rather uniform cells.

Petioles have been successfully rooted and also have been grafted into tubers, when they grew into abnormally thick organs. In such cases the normally flattened collateral bundle of the petiole tends to become concentric, and the petiole comes to resemble a stem in structure and function. The change is due to the formation of new cambium by division in cells of the ground parenchyma, either extending out from the edges of the original cambium and curving round till the two sides unite, or beginning in groups of parenchyma cells opposite the central part of the collateral bundle, the process then spreading laterally and curving down to join the original cambium on each side. Xylem forms on the inside of the new cambium and phloem outside. Ultimately a compact woody cylinder forms, which may be twenty times the area in transverse section of the normal petiqle bundle. A small-celled phloem surrounds the wood, and all the elements that are formed outside the original bundle are short, as they are limited by the diametre of the ground tissue cells from which they are derived.

It has, indeed, been shown experimentally that all living plant tissues have potentialities of hypertrophic and hyperplastic reaction to 'physiological' stimuli in the absence of wounding or of any parasite. In the hyperplastic tissues differentiation to other tissue forms can follow, but there is a tendency for them, at first at least, to show less differentiation in the elements of which they are composed than is found in similarly situated normal tissues. Naturally, cells that have not finished growth and differentiation react the more readily to such stimuli, but old cells can also be affected, either directly or after first dividing and thereby becoming, so to speak, rejuvenated.

Even cells that have developed thickened walls, as collenchyma, can enlarge and divide, or those in the early stages of lignification can undergo delignification and resume activity. In the hyperplastic tissue a cell, perhaps only one of a group formed from a single mother-cell, may become a thick-walled element such as a sclereid, or develop into a little island of phloem or a tracheid; or a meristem may appear, with all its powers of tissue production. Similarly hypoplastic modifications can result from under-development brought about by inadequate or unbalanced nutrition or other unfavourable conditions.

TISSUE MODIFICATIONS CAUSED BY PARASTITES

The tissue-changes that may result from parasitic attack on plants, remarkable though they may sometimes be, resemble in many respects those that have been described above, allowance being made for the fact that the stimulus may be more localised and may continue for a longer time than that caused, for instance, by wounding. It seems very doubtful whether there is any essential difference between the modifications of tissues and organs found in galls due to parasites and those that may occur in wound calluses or the various other types of reaction to wound injury or physiological causes.

Hypoplastic modifications are seen in the sterility caused by some parasites, in the small size.of the leaves in trees attacked by certain virus and other diseases and in inhibition of tissue differentiation as a result of parasitic attack. An interesting example of the latter is the prevention of the formation of an absciss layer in cherry leaves attacked by *Gnomonia erythrostoma*, with the result that they do not fall in the autumn but remain hanging on the tree and serve to infect the new leaves in the following spring.

Host-tissue reactions to parasitic attack by the formation of wound gum in woody tissues and wound cork in all living parts resemble those due to simple wounding in their main characteristics. The band of reactionary cork formed around a leaf infection can sometimes be seen in thin leaves as a narrow dark circle on holding them up to the light. The cork may be effective in checking the parasite or may fail. A common result of its formation in leaves is the necrosis and dropping out of the surrounded area, producing what is known as a 'shothole' effect.

It has been already mentioned that there is a tendency for the tissues of reaction to non-parasitic stimuli to consist at first of elements resembling one another, as, for instance, in calluses and intumescences, in the repaired epidermis, palisade and spongy parenchyma at the edge of a leaf wound, or in the thin-walled derivates of the cambium on the xylem side in the thickened shoots of kohlrabi prevented from flowering in the second year. This tendency is still more marked in many galls and has led to the use of the term' gall' or 'tumour' tissue to characterize the masses of soft rounded cells, often with few or no intercellular spaces, that may develop as a result of the presence of a parasite.

Galls are swellings on plant organs due to a stimulating or irritant action of parasites such as fungi, bacteria or insects on the tissues. In a few cases they are caused by viruses. They may vary from tiny swellings involving a few cells of the epidermis as in the early stages of wart disease of potato to the soft rounded tumours several inches across of maize smut, or the large woody masses formed sometimes in crown gall or in certain rusts on Acacia. The term gall is sometimes extended to include organic outgrowths or proliferations, such as the leafy organs formed in cereal ears attacked by *Sclerospora* and also to cover cases of uniform enlargement of an organ, as in flowers of some *Cruciferae* infected by *Cystopus candidus,* or the 'bladder plums' produced by *Taphrina pruni.*

Galls due to Tissue Swellings*: *'Histoid' Galls

Some of the simplest galls are caused by species of Synchytrium. The fungus may be confined to a single epidermal cell which generally becomes much enlarged. Neighbouring cells also react, usually hyperplastically, and the host cell becomes surrounded by a sheath which, as in *S. taraxaci* on the dandelion, may be derived from both epidermal and subepidermal cells.

The bordering cells in this little gall elongate tangentially to the host cell and divide by walls at right angles to the latter. A small-celled sheath of two or three layers in depth is thus formed around the enlarged cell containing the fungus. No intercellular spaces occur in the sheath nor is there any detectable difference between the derivatives of the epidermis and mesophyll that constitute it; there appears to be also no tendency to form cork, perhaps because of the short duration of active stimulation.

Another very simple epidermal gall is caused on poplar leaves by *Taphrina populina,* syn. *T. aurea,* the mycelium of which is situated between the cuticle and epidermis and for a short distance down between the cells of the latter. These cells elongate to two or three times the normal length and may divide once or twice, the result of their uneven growth in the developing leaf being the formation of a small bulge on one side of the leaf and a corresponding depression on the other.

Another leaf gall of a simple type is that formed on apple leaves by the aecidial stage of *Gymnosporangium juniperi-virginianae* in America. The affected part of the leaf is swollen to two or three times its normal thickness by a great increase in height of the spongy mesophyll cells, the palisade being little altered. The elongated cells divide by transverse or oblique walls and the intercellular spaces are lost. In the thickened leaves produced on the shepherd's purse by infection with Cystopus candidus, the mesophyll is composed wholly of large round cells, distinction between palisade and spongy cells being lost.

Taphrina deformans also thickens the peach leaves that it attacks, and causes the palisade cells to multiply and lose their normal shape, becoming almost isodiametric in the outer layer, where they are hard to distinguish from epidermal cells. *Exobasidium oxycocci* makes all the mesophyll into a uniform type of parenchyma without intercellular spaces in *Vaccinium macrocarpon,* one of the cranberries, but in the thickening caused by *E. vaccinii in V. vitisidaea* leaves, the palisade is little altered and the swelling is due to the development of roundish or polygonal cells with no intercellular spaces in the spongy parenchyma. A simple stem gall is the common one on goutweed or bishopweed caused by *Protomyces macrosporus,* where there is a hyperplasy of the outer cortical cells, restricted to those in the neighbourhood of the intercellular hyphae. The easiest way to locate the hyphae of this fungus in the early stages of infection is to look for a fully formed cortical cell in process of division.

Somewhat more complex are the small corky swellings that constitute the disease known as scab on the leaves and twigs of citrus plants. The parasite, a conidial stage of the fungus *Elsinoe fawcetti,* causes in attacks on the under surface of the leaf a necrosis of a few superficial cells, and provokes a hyperplasy of the underlying spongy parenchyma. The new cell walls are roughly parallel to the necrosed surface, and the reaction may extend through the leaf and cause some of the palisade cells to divide. Cells near the lesion elongate considerably and divide by several cross walls. A band of the new cells thus formed becomes a cork cambium. This extends up to the epidermis all round the lesion, and cuts off cork on the side next the latter and a little phelloderm towards the sound tissues. All the primary cell walls of the hyperplastic area are thickened, and intercellular spaces are much reduced.

A good example of a gall restricted to the vascular region is that produced by *Sorosphaera veronicae* on *Veronica chamaedrys,* the germander speedwell. The swelling starts in the procambial region of the shoot, from the point behind the apical meristem where the xylem is just beginning to form, and extends backwards without involving the cortex or pith, and sometimes outwards along a young leaf bundle. Spiral vessels can be found throughout its length but do not occupy any fixed position in relation to the mass of gall tissue, being sometimes on its inside towards the pith, sometimes on its outside. Towards the cortex, sometimes in the middle. This shows that the gall tissue is formed not only from the cambium, but from cambial derivatives lying around the protoxylem vessels. I t consists of more or less polygonal cells of a very uniform type, the only other elements found in the swollen part being the few xylem elements that were already differentiated before the parasite reached the area, and some scattered groups of elongated cells. The parasite, an intracellular one, does not appear to occur in all the tumour cells, but the hypertrophy and hyperplasy are general throughout the region involved. The cortex and pith usually undergo no histological modifications, though there are indications that, in later stages, the parasite may reach the inner cortex by passage from within through the endodermallayer.

In the finger-and-toe or club-root disease of turnips, cabbage and other crucifers, the parasite Plasmodiophora brassicae is again mainly confined to the central cylinder. In inoculations on cabbage stems the fungus reaches the cambium fairly soon and induces a marked hyperplasy in the cells in this region. The elements cut off on the xylem side remain undifferentiated and again rapidly divide. In older stems the medullary rays are chiefly infected and show great hypertrophy and hyperplasy. Except at the point of original invasion, the phloem and secondary cortex only become infected outwards from the cambium. In roots the bulk of the hyperplasy is in the cambium and the phloem initials adjoining it; the new cells in the position of the phloem are thin-walled and accumulate starch. The increase in size of the medullary rays may divide the vascular ring into widely separated bundles which ultimately may become much distorted by the uncontrolled growth in the rays. New tracheidal tissue develops in the rays. The changes that occur in the vascular region in this disease show a certain similarity to those already mentioned as occurring in allied plants as a result of physiological processes consequent on the prevention of flowering.

In various other types of gall, insect as well as fungal, the tendency already mentioned for the cambium, when stimulated to abnormal activity, to cut off non-lignified parenchyma without fibres and with fewer and thinner-walled tracheids on the xylem side, and secondary parenchymatous phloem outside is strongly marked.

In some of those that cause the medullary rays to enlarge, so as to separate the vascular ring into isolated bundles, these tend to become bordered by a cambium forming in the ray cells and in the region of the protoxylem, very much as was found in the thickening due to prevention of flowering in kohlrabi. The underlying causes of these processes are not known, but the similarities are sufficient to suggest that whether the stimulus is a fungus or an insect or purely, physiological', the response follows along certain broadly defined lines.

In some fungal and bacterial galls all or nearly all the tissues may become involved in the swelling. The thickened axial shoots that are the most prominent feature of the destructive witches' broom disease of cacao in the West Indies and South America, due to *Marasmius perniciosus*, are good examples of this. As fuany as a thousand of these may be found on a tree. The swelling may involve every tissue from epidermis to pith, sometimes almost equally, each being two or three times the normal thickness, sometimes mainly in the cortex or the xylem. Most of the thickening in the xylem seems to be from cambial activity, but in the other tissues it is mainly due to division of parenchyma in situ.

The pericyc1ic region and deeper tissues may proliferate to such a degree that the cortex is crushed, or may become so parenchymatous that it is almost impossible to define the inner limits of the cortex, though scattered groups of pericyc1ic fibres can still be distinguished. All the parenchyma cells, except those nearest the epidermis, assume the same shape, elongated radially and with thin walls. The xylem vessels are broader and thinner than usual and the part of the xylem near the cambium is largely parenchymatous.

The fibres become at first septate, then reduced in numbers, and finally are not formed. As the thickened shoot grows in length, tissue differentiation is decreased. In the swollen leaves the distinction between palisade and spongy parenchyma is lost. The 'brooms' are short lived and remain soft and fleshy, cork being rarely found in them. The huge *Uromycladium* galls formed on species of Acacia in Australasia start usually by a proliferation in the phloem, infection beginning generally before cork has developed. The outer tissues are deformed and ruptured by pressure from within, due both to expansion of the phloem and pericycle and to the production of wound wood composed of unlignified rows of tracheids, with a few wide vessels and an excess of parenchyma.

The identity of the original cambium is lost at an early stage, and 13;ter it becomes impossible to distinguish clearly between phloem, cambium and outer xylem. Cork may form in the pericyclic region or the outer cells may continue to proliferate as hyperhydric tissue. Tracheids develop in any of the tissues that take part in the formation of the gall.

No plant gall has been so completely studied as crown gall caused by *Bacterium tumefaciens*, because of its intrinsic interest, its wide range of host plants, some of considerable economic importance, and because of suggested analogies between it and cancerous growths in man and animals.

The organism which causes crown gall is one which is not ordinarily provided with enzymes capable of dissolving the polysaccharides of cell walls or starch, so that it does not form cavities or penetrate into the interior of the plant cells or induce soft rotting. On the other hand, it has marked powers of stimulating the cells in its immediate vicinity to divide and form hyperplastic masses of tumour tissue. When it is inoculated into the tissues of herbaceous stems that are still growing, it may occur in the intercellular spaces of considerable lengths of stem formed subsequently to the inoculation and it may also spread actively or passively in the fully grown internodes, along the xylem or where, as in the pith or cortex, intercellular spaces are in erotical communication with one another for some distance longitudinally.

In these cases it may give rise to continuous strands of tumour tissue formed by the division of bordering cells, or it may only cause visible hyperplasy here and there along the course of its extension. It is not uncommon to find, following inoculations of this type towards the top of the stem, closed vascular rings, spheres or cylinders in the pith and cortex. When these are found in the pith the orientation of the tissues is frequently inverted, the phloem being inside and the xylem outside the cambium.

In the cortex and petiole the orientation is normal, and in the latter situation the stimulation from the parasite may lead to the formation of a closed concentric bundle by division of the ground cells of the petiole outside the leaf trace, just as occurs when a petiole is rooted as described above. The centre of the new formation is a mass of tumour tissue in which cells of the ground parenchyma not transformed into tumour cells may sometimes be seen. In some of these infected petioles the organism has been found to occur in the protoxylem, and the meristematic divisions in the ground tissue begin opposite this point.

Tumour strands in the outer cortex resulting from inoculations of *Bact. tumefaciens* in this region have been found to arise from extension of the bacteria along intercellular spaces running longitudinally in the stem or drawn out by "rapid elongation of young parts, the cells bordering the infected spaces being stimulated to active division. A secondary gall has been found to develop at some distance from the point of inoculation, and in this a closed vascular sphere may occur, having no vascular connection with the central cylinder of the stem. In some pith strands the cells bordering the first-formed tumour cells may show only hypertrophy but in others a vascular cylinder develops, with xylem outside.

The vascular tissue may form around a core of tumour cells in a part only of these strands; in strands in the pith and cortex it develops independently of the original cambium and usually at some distance from it. The inverse arrangement of the bundles thus formed in the pith resembles that found normally in many families of plants possessing medullary bundles. Sometimes also, both in normal pith and in the protoxylem of inoculated plants, only a little nest of phloem occurs without cambium or xylem, and it is interesting to note that a similar formation has been found in the middle part of the protoxylem parenchyma in kohlrabi plants prevented from flowering. The exciting cause of these inverted normal or induced formations appears to be unknown, though attempts have been made to explain them.

When crown gall results from inoculation into the deeper layers of a herbaceous stem the tumour tissue may appear to burst through the central cylinder and cortex to the outside. This appearance is due to a progressive hyperplasy of the ray and cortex cells, the outer layer of the latter being sometimes crushed by pressure from below. In such erupting galls it is easy to follow the conversion of the normal cells into gall tissue by multiple division. In the tomato, the gall cells thus formed become smaller and smaller, by division, for some weeks, then enlarge somewhat; an average normal cell of the cortex may be nearly 5000 sq. μ in area, and this may fall by division to 166 μ after about a month and rise to over 400 μ in about two months.

The tumour tissue from secondary tumours in herbaceous plants appears to be able to continue growth after removal from the infected plant. It still goes on growing when, after several' generations', bacteria cannot be isolated from the mass of tumour cells and the ground-up mass is incapable of infecting healthy plants. The apparently sterile tumour mass can be implanted by grafting into a healthy sunflower and will continue to grow into a tumour resembling a typical crown gall on the plant, reaching sometimes a diametre of I cm. after seven weeks. Work at East Mailing has shown that similar tumours can be produced by the growth-promoting substance indolebutyric acid, while in America it was found that indoleacetic acid causes both primary and secondary tumours, the latter at some distance from the treated site.

Though it has been found that *B. tumefaciens* can form indoleacetic acid in media containing tryptophane or other nitrogen compounds of high molecular weight, and the accumulated evidence indicates that a diffusible substance is concerned in the development of the tumours, it is not yet clear what the substance is or whether it is a product of the host cells or of the metabolism of the bacteria, or even if it is one of the recognised growth promoting substances, such as indoleacetic acid, at all.

In woody stems such as those of the raspberry or blackberry, in which crown gall is not uncommon, the galls may appear as multiple outgrowths. In these infections the swelling may at first be confined to the region of the pericycle outside the fibre bundles. There may be no nests of tumour tissue, as in.the herbaceous stem galls, the swelling consisting of thin walled radially elongated parenchyma in rows and with scanty contents.

The multiplying cells lie below the cork, in the pericycle, which form in layers in the neighbourhood of the endodermal region. At this stage the new tissue may be regarded as a parenchymatous derivation of the actively dividing pericycle, but the reaction soon extends down between the fibre bundles and involves the tissues between them and the outer part of the phloem. The growth in this position bends up the pericyclic fibres and separates groups of them widely from one another. Here and there, usually not over the whole arc underlying the gall, the cambium is stimulated tq form wound wood, which is sharply differentiated from the normal xylem by a scarcity of fibres, the thinner walls and wider lumina of the tracheids, and the absence of large vessels.

In the tumour mass outside the cambium single tracheids or tracheid groups are formed by differentiation in individual tumour cells, just as occurs in an ordinary callus outgrowth and, as in callus, the tracheid islands may become bordered by a meristem producing further tracheids on one side and phloem or gall parenchyma on the other. The gall often becomes highly vascularised in this manner and some of the strands may reach the original central cylinder.

The surface of the gall is usually a loose mass of brown collapsed cells with weakly lignified walls, resembling the hyperhydric tissue frequently formed on the sur face of a callus. There is, indeed, a fairly close histological analogy between crown' gall and wound callus of the type found causing knots on apple grafts.

There is little reason for the supposed analogies between this plant gall and the malignant growths that form animal cancers. As will have been seen, the processes involved in its formation are not essentially different from those found in other plant galls. Similarly, the leafy outgrowths that sometimes occur on the galls, and that have been termed embryomata from a suggested resemblance to one type of cancer, are no doubt to be likened to callus shoots.

Organoid Galls

There is another type of structural modification sometimes produced in plants as a result of the presence of a parasite in the tissues to which, though the changes are morphological rather than histological, the term' gall' is often applied. These, organoid' galls, as distinct from the' histoid' type already discussed, result especially from systemic in actions with obligate parasites of the kind that do not kill the invaded plant parts at an early stage of parasitism. They are characterized by changes in the morphological plan of the shoots or other organs affected, or in the rhythm of their development.

Marked changes in symmetry and habit may be found in plants affected by some rusts and mildews. A good example is the change from the normal short unbranched stem, with radical leaves, of *Launea asplenifolia,* a common Indian weed, to an elongated much branched axis, with cauline leaves, when infected with *Puccinia butleri,* a systemic rust. Stem elongation is caused by *Ustilago hypodytis* on *Elymus arenarius* and *Sclerospora sacchari.* on sugar cane and the number of internodes and leaves is increased by various rusts, e.g. *Uromyces pisi* on *Euphorbia cyparissias* and by *Gibberella fujikuroi* in the 'bakanae' disease of rice in Japan, where it has been traced to the production of a crystalline active principle gibberellin; shortening of internodes occurs in rye attacked by *Urocystis occulta.*

The symmetry of the leaves is altered in some diseases. They may be changed from simple to irregularly lobed in Berberis buxifolia attacked by Aecidium jacobsthalii in South America. In the' reversion' disease of black currants one of the most useful aids to diagnosis is the change in the venation and, especially, the lobation of the leaves of affected plants. The reverted leaf may have only three main veins, diverging from the top of the petiole to the extremity of three main leaf lobes, instead of the usual five of each. Lateral sub-veins from the middle main vein are reduced tq about half the normal number, but they serve to supply nearly all the teeth on the leaf margin, whereas the 10 to 16 sub-veins of the normal leaf do not supply more than about half the teeth of the apical lobe. The reverted leaf, best seen on wall grown shoots in May and June, is smaller, narrower, flatter at the base, coarser and of a deeper green than normal.

An extreme example of the opposite tendency is to be found, as the name implies, in the 'fern leaf' type of tomato mosaic disease while some forms of tobacco virus infection can cause a twisting of the leaf so that the under surface faces upwards. Another remarkable effect of virus infection on leaves is the formation of 'enations "green cup-shaped or blade-shaped protusions, generally on the under side of the leaf. The very remarkable leafy outgrowths from the leaves of bracken caused by the fungus Taphrina laurencia in Ceylon are probably due to processes similar to those causing enations.

Perhaps the most profound modifications of the type under discussion are caused in inflorescences and single flowers by the action of certain fungi and viruses. One of the commonest of these changes is the inducement of 'phyllody' or the transformation of flower parts into leafy organs. This is found in several virus diseases of tomato, sandalwood and other plants, in the flower of the Japanese plum *Prunus mume,* infected by *Caeoma makinoi,* and in certain crucifers attacked by *Cystopus candidus.*

Single flowers may be doubled by the development of additional petals or their symmetry may be changed from the regular to the irregular plan and vice versa; in *Matricaria inodora* attacked by *Peronospora radii* ligulate flowers may be found in the centre of the disc and tubular flowers at the margin). The rudimentary stamens in pistillate flowers of Lychnis dioica may become fully developed when attacked by *Ustilago violacea,* except that the pollen is replaced by spores, while in the contrary direction the maize' tassel' when infected by *U. zeae* may bear female or bisexual flowers, or even normal grain at its base. Some rusts prevent flowering of the host when there is a perennial mycelium extending from the root-stock to the growing point but permit normal flowering when the infection is local.

The effects produced on the inflorescence of certain crucifers by *Cystopus candidus* may be very intense. The flower stalk and axis of the inflorescence may be enormously thickened while the floral parts are wholly or in part fleshy, green or violet and persist instead of falling early. The petals may resemble sepals and the stamens become leaf-like or occasionally like carpels. The latter may be open instead of united into an ovary, while the ovules and pollen grains may be atrophied, thus causing sterility; sometimes the pistil may be much swollen into a conical thickwalled sac containing sterile ovules while the anthers may have supplementary pollen sacs or become like carpels by the development of stigma-like structures at the tip and rudimentary ovules along the margin.

Occasionally the flowers revert to the primitive spiral insertion of their parts. Chlorophyll and starch may be copious even in the petals and stamens; hairs may develop on the anthers, and stomata in the inner epidermis of the enlarged ovary. In this case the tendency to a lessening of differentiation in the tissues of the swollen parts, which is marked, is carried further to a tendency to reduce the differentiation of organs.

One of the most striking cases of virescence of the inflorescence accompanied by leafy modifications of the floral parts is the 'green ear' disease of millets in India, Africa, Japan, the United States, and elsewhere, caused by *Sclerospora graminicola.* As the name suggests, the chief symptom of the disease in the commonly affected bullrush or pearl millet, *Pennisetum typhoideum,* is the transformation of the solid spicate ear wholly or in part into a loose green head, composed of a mass of small twisted leaves.

This is due to the replacement of the upper part of the axis of each floret, which normally forms the grain, by a short leafy shoot. The bristles below the glumes are enlarged and contorted and the fertile glumes enlarged and sometimes turned green. The stamens are often converted into minute leaf-like organs with a distinct division between blade and sheath. When not so transformed, they may occur as brown pointed bodies with no trace of the anthers, or if the anther is formed, its size relative to that of the filament may be much altered and pollen may be absent.

The pistil often undergoesedian proliferation, being replaced by a small leafy shoot, but sometimes by a minute branched axis with undeveloped buds, or by a simple, hollow, hornlike growth, which may be a leaf united along its edges. Sometimes two spikelets arise from the pedicel instead of the usual one, and the number of florets on a spikelet in infected ears may be increased from two to three or four. On the other hand, oats infected by *Sclerospora macrospora* in Mississipi bore only a single seed-like structure in the glumes instead of the normal two close together. The median floral proliferation, which is the main characteristic of the' green ear' cereal diseases, occurs also in certain virus diseases.

A remarkable instance of structural modification is caused by an organism first isolated in the United States from fasciated sweet peas. Later, this organism, *Bacterium fasdans*, was found to be parasitic in a number of garden and other plants in England and to cause leafy and fasciated outgrowths from the buds at the crown of the root-stock and at the stem nodes, when inoculated in these sites. In some plants the leafy shoots from the base are so numerous as to appear like moss.- Bud inoculations on seedlings cause galls at the growing point and in leafaxils, with a complete check of normal growth; in other plants fasciated or cauliflower-like shoots are produced. Proliferation is confined to the bud tissues, inoculations elsewhere on leaves, stems and roots producing no effect. Amongst the 25 genera from which *B. fascians* has been isolated are peas, beans, strawberries, chrysanthemums, petunias, and asparagus.

In strawberries a high proportion of inoculated seedlings of the St.-Jan variety developed leafy and floral abnormalities resembling those caused by cauliflower disease, but the inoculation of 'runner' plants gave inconclusive results.

The term witches-' broom is applied by the country people in some parts of Europe to irregular tufts of crowded twigs found in trees and shrubs. They have, as indeed have several of the' organoid' type of galls above-mentioned, some of the characters of 'histoid' galls, for the tissues that take part in their formation are often swollen. Their chief character, however, is the production of short twigs in irregular groups, often all turned upwards and with small leaves.

When due to fungi, rusts and species of Exoascaceae are mainly concerned; the fungus passes into the new shoots produced each year, and often fructifies on them. When spore production is finished, the leaves and sometimes the younger twigs die off. Well-known examples are the brooms produced by Taphrina cerasi on cherries and those on silver fir due to Peridermium elatinum. Mites and insects sometimes cause similar growths, and some brooms on pine and spruce are not caused by parasites, but appear to be due to an innate heritable tendency of the tree.

In galls, whether 'histoid' or 'organoid', with rare exceptions, there is nothing new in the tissues or organs of reaction, but merely a disorganisation or an intensification or inhibition of normal processes of tissue and organ formation and differentiation. The parasite can do no more than call out or suppress powers which the cells and tissues already possess, powers which plants can ordinarily make use of in dealing with such accidents as wounding or in meeting extremes of environmental or nutritional or growth modifying conditions. Indeed, most of the modifications in tissues and organs caused by parasites can be paralleled by those brought about by physiological processes without the intervention of a parasite, or by examples selected from the extensive literature of plant teratology.

In the 'quercina' virus disease of the thorn apple *Datura stramonium* the spines on the capsules are eliminated, the stigmatic surface of the pistil is elongated, the tubular corolla becomes one with separate petals, and the leaf margins are eroded. All these modifications have been recently found in a hereditary mutation of the, same host. In speculating on the causes that may underlie some of these modifications, especially those of the type found in organoid galls, it is impossible to avoid the suggestion that they may be due to dislocation in the production or movement of growth controlling substances, much light on the existence and properties of which has been obtained in recent years. Little knowledge, however, is as yet available as to the part they play in stimulating gall formation though, as already mentioned, they seem to be concerned in the production of crown-gall tumours.

It is worth mentioning that phytohormones are formed in abundance by some fungi and may have a similar effect to those in a crude lentil extract in promoting growth in *Nematospora* and fructification in many fungi. It is not only the meristems that have the ability to respond to stimulation by the development of new tissues or by modifications in the products of their activity: other living tissues can playa part not less important.

No doubt meristems respond very readily, and there may be indications that the pericyclic region is easily stimulated into activity in some plants, but the examples given above show that this sensitiveness must not be given undue importance. The generalisation has been made that in all living plant cells there slumbers the potentiality for the development of all histological characters that appertain to the particular species. The study of galls gives point to this remark.

EFFECTS ON CELLS AND CELL MEMBRANES

So far the pathological modifications discussed have been chiefly those affecting tissues and organs of the plant. The changes in single cells and cell membranes brought about by the presence of a parasite have been only incidentally referred to both in this chapter and in earlier ones, but it will have been inferred that they fall into certain main groups according as the parasite is a cause of cellular necrosis or is one which stimulates the cell to divide and become more or less meristematic, or even lives in equilibrium with the host, causing no appreciable change in the earlier stages of its development.

CELL CONTENTS

The cell may be killed without reacting against the attack. There may be no increase in size or change in the nature of the contents, except such as results from their death and consumption by the parasite. The whole cell contents may collect towards the centre in a grumous mass, which turns brown apparently from the oxidation of substances like tannin present in the cell, and breaks up into amorphous granules often with the formation of refractive brown globules; these have been mistaken for parasites on several occasions. Sometimes the liquid of the cell is rapidly removed, and the ordinary effect of wilting from drought is produced.

Starch may be hydrolysed, or may be left unaltered especially when the cell has been rapidly dried out. Several of the potato tuber rotting species of Fusarium can hydrolyse gelatinous starch, but have no effect on the' unbroken starch grains of the tuber. Chlorophyll may be reduced-leading to pallor of the infected area or destroyed, when browning usually follows. Or it may be increased or preserved to form green islands after the rest of the green tissue withers. In some rusts there is evidence that not only is the chlorophyll preserved but the chloroplasts are stimulated to increased activity by the fungus.

A development of red or purple pigment due to the formation of anthocyanin often results from infection by rusts, *Exoascaceae* and other fungi, especially on the side towards the sun. This is also a 'frequent occurrence in the abnormal stems, petioles and leaves produced by systemic rusts and other parasites such as *Bacterium fascians*. The common *Oidium euonymi-japonici* causes large red blotches from this cause on *Euonymus* leaves in the spring, which extend beyond the area bearing the haustoria of the fungus. The nucleus may be enlarged or become lobed or even fragmented, as in cells of the shepherd's purse attacked by *Cystopus candidus*. Calcium oxalate crystals may be increased in apple scab, or reduced in buckthorn infected with the aecidial stage of *Puccinia coronata.* Infection is reported often to stimulate the activity of the mitochondria, and also to cause an accumulation of phenolic compounds in the cell vacuoles.

These have been thought to exercise a protective action against parasites. The distastefulness of uncongenial or resistant hosts to some fungi is probably due to the presence of substances in the cells; but it is a local reaction and not comparable with the formation of true antibodies such as play' so important a part in the protection of the animal body against disease.

CELL MEMBRANES

Action by the parasite and reaction by the host are more obvious in the cell membranes than in the cell contents. It has already been mentioned that the parasite may penetrate cell walls by pressure or by the solvent action of enzymes, pressure being chiefly used against cutin and suberin but also sometimes against internal walls devoid of these substances, while the pectin and cellulose membranes of the softer interior tissues are usually dissolved by enzymic action. Action on cellulose membrane~ has been, perhaps, most fully studied in certain leaf and twig parasites which occupy chiefly the subcuticular and epidermal walls and in fungi which attack woody tissue; action on the lignin is best seen in woody tissues; while the soft rot which results from pectin dissolution has been most closely studied in tubers, fruits and other storage organs.

Cellulose is the most important constituent of plant membranes, forming the bulk of the walls of the softer tissues and of all young parts that have passed the pectic stage. Most parasitic endophytes are capable of dissolving cellulose by means of cellulose enzymes. Sometimes the power is restricted to certain organs or stages in development of the parasite, as in a number of *Peronosporaceae,* smuts, and the like that are intracellular in or near the epidermis in the early stages of penetration but are subsequently confined to the.intercellular spaces; pressure, however, may be involved also in some of these last. Cellulose and other membranes may be penetrated by a hole which does not subsequently enlarge, presumably because in such cases the enzymes involved are.released only from the tips of the hyphae.

In the subcuticular parasites whose vegetative thallus is mainly just below the cuticle or in the outer surface layers of the epidermal cell walls, entry into the host is effected by a fine infection hypha; this penetrates the cuticle and enlarges below into a mycelium which does not gain access to the lumen of the underlying epidermal cell. Some seem to dissolve their way in the cellulose part of the wall others to force a passage by mechanical means. The epidermal walls are often impregnated with cutin, cutinized walls sometimes extending to the radial and inner sides of the cells; mechanical force may be required in passing along these, since cutindissolving enzymes have not been found.

But in most fungi of this type there are evident signs of dissolution of the cellulose and little doubt that the fungus gets part of its carbohydrate nutrition from this source. In practically all subcuticular fungi -there are, at the same time, signs of depletion of the underlying cell contents, modifications in the chloroplasts, desiccation and browning, often accompanied by a formation of wound cork deeper in, so that there is evidently a passage of substances from the fungus to the host cell and, presumably, a transfer of nutrient in the reverse direction. The vegetative thallus, which often extends down to some extent in the radial walls of the epidermis, eventually is able to complete its development and form spores, set free by rupture of the cuticle.

Another group of fungi which consume the cellulose of the cell membranes comprises those responsible for the brown rots of wood. This is a large group containing both parasites and saprophytes but all characterized by attack restricted more or less completely to the cellulose, pentosans, and hexosans of the wood, and all increasing the acidity and the solubility in alkali of the products of decay. Even in extremely decayed wood, however, all the cellulose is not necessarily destroyed and part of the pentosans may also be left. Lignin is little affected in this type of rot, the enzymes concerned in which seem to be those which effect hydrolysis. All the fungi involved can use cellulose as a source of carbohydrate nutrition; tests at the Forest Products Research Laboratory at Princes Risborough have shown that Trarnetes serialis can live with chemically pure Sitka spruce cellulose as its sole carbohydrate nutrient. The cellulose is usually converted into glucose before consumption.

All the fungi in this group can also use the hexose and pentose sugars that result from the hydrolysis of hexosans and pentosans. Easily hydrolysable hexosans such as mannan and galactan may be amongst the earliest substances to be consumed, though more usually the pentosans are the first to disappear.

Amongst the saprophytic species affecting wood in this way the best known is *Merulius lacrymans*, the chief cause of dry rot in structural timber in temperate, climates. This fungus dissolves much of the cellulose in six months during which time it can cause a loss of 45 per cent by weight of the wood, rising to about 7° per cent in two years. Like many other wood destroyrs it forms much water from the wood, which drips away and also provides the moisture necessary for its further progress into dry timber.

There is little or no delignification, the residue consisting of a rather homogeneous lignin body apparently similar to that yielded by sound wood on treatment with caustic alkali under pressure. A certain amount of the other constituents of the wood, more or less modified are also left. The methods content of the decayed wood may be increased by as much as 77 per cent and uronic acid is also increased. The final product of the decay is considered to be closely allied to humus. *Coniophora cerebella* can also reduce the weight of the timber by about 7° per cent and leaves a humuslike residue. In the brown rot caused by Poria vaillantii the loss of weight in Scots pine sapwood may be about 55 per cent after two years and in spruce wood attacked by this fungus all the cellulose may be gone in six months.

Trametes serialis is one of the most important sources of decay of coniferous timber in western North America, causing a cubical rot of redwood and a butt rot of Sitka spruce, Douglas fir, and other softwoods. It causes an acid hydrolysis of Sitka spruce which alters the chemical composition, increases alkali solubility and reduces the strength of the wood before any loss of weight can be detected.

The loss in mechanical strength may exceed 15 per cent in the first fortnight, whereas it may take four or five weeks before any significant loss of weight occurs. The wood is composed, as to about 99' I per cent of its weight, of cellulose, lignin, pentosans not in the cellulose, and watersoluble material, and this fraction was found to lose only about 4 per cent in three or four months, the loss falling on the cellulose and pentosans. The rise in alkali solubility runs parallel to loss of strength, which may increase steadily to about 80 per cent.; then it slows down. In the early stage there is little change in the appearance of the wood, and aeroplane timber which has failed under test may show, on microscopic examination, hyphae well in advance of any visible sign of decay. With *Polyporus schweinittzii* it causes the condition known as 'dote' in Sitka spruce, one of the chief timbers used in aeroplane construction.

Other active brown rot fungi which have little effect on lignin include *Fomes pinicola,* which is sometimes parasitic, but very rare in Britain, *Lenzites trabea* and *L. abietina. L. sepiaria,* which is a fairly common cause of decay of pine timbers, is found in pit props, can reduce the wood of *Pinus taeda* and *P. caribaea* to a brown crumbling mass and has an exceptional armament of enzymes, and *Lentinus lepideus,* which eventually may cause a slight loss of lignin owing to hydrolysis of the methoxyl groups in it, and which may, like *F. pinicola,* cause large holes in the. wall when it passes from cell to cell.

The other main group of fungi causing decay in standing or fallen timber consists of those responsible for the so-called white rots. The white rots are more important to the forester as a rule than the brown rots, since they comprise a number of the most serious parasites of the wood of trees. Their action is not limited to hydrolysis for they possess also oxidizers enzymes, in contrast to the brown rots, many of which show reductase activity. They attack all the major components of the cell walls of the wood, but there is no uniformity of sequence or proportion in their destruction of these. Some attack the chief components of the wood simultaneously, others cause delignification before the cellulose is depleted, others again show a delayed delignification. The variations are in part due to differences in the time when one or another type of enzyme becomes active, and the course and speed of the secondary reactions. Where de lignification precedes cellulose decomposition, it has been suggested in certain cases that an oxidase from the fungus acts first on lignin and pentosans to produce acid and that the oxidase with acid can complete the process involving the depletion of the cellulose. The lignin destroying fungi can digest lignin presumably after decomposing the aromatic content and the aliphatic carbohydrates that constitute its two major groups of compounds.

Polystictus versicolour is a very good example of the white rotting fungi, for it can so completely destroy beech wood as to leave only a fraction of I per cent after two years. It attacks the ray cells and wood parenchyma first, reducing the lignin, pentosans not in the cellulose, cold-water soluble and I per cent. alkali soluble compounds, before the cellulose and its associated pentosans and the hot-water soluble fractions are diminished. It also bleaches the wall pigments. Like many other white rotting fungi, the I per cent caustic soda solubility of the pentosans not in the cellulose diminishes as the decay progresses, so that there is ultimately no marked increase in the total alkali solubility as compared with sound wood. Loss in strength exceeds loss in weight.

Trametes pini causes a red rot in some hosts, but the white or pocketrot type of decay is eventually found and is due to a progressive delignification working outwards from the tertiary layer of the membrane to the middle lamella; cellulose is attacked later. Like *Fomes annosus* it may dissolve the middle lamella, causing the cells to fall apart, before the cellulose is completely destroyed. Trametes suaveolens follows the same course from within outward in its action on the wall in willows and poplars, but cellulose is depleted before lignin. *Ganoderma curtisii* causes a white basal rot of hardwoods in the South-East of the United States, destroying the middle lamellae of the vessels and large ray cells, delignifying these and corroding the bordered pits into large holes. Stereumfrustulatum causes white pocket-rot of oak byan early concentration of delignifying activity in oval areas separated by firm woody tissue.

The pockets at first contain cellulose with some hyphae but are eventually empty. All the wood-decaying species of *Stereum* act in much the same way, the white areas being merely places where the lignin decomposition has proceeded most rapidly so that a higher proportion of white cellulose is left, but some of the cellulose is consumed early and eventually most of it disappears so that there is no real selective action on lignin apart from the time factor.

Armillaria mellea causes a decay which has some of the characters of both the brown and the white but can best be included with the latter, with which it shares the capacity to produce both hydrolysing and oxidising enzymes and failure to increase appreciably the total alkali solubility of the wood; at all stages this solubility is greater in the brown than in the white rots. *A. mellea* begins by attacking the cellulose and associated pentosans but it depletes the lignin also, causing a soft, wet white rot, while in advanced stages of decay it may cause coniferous wood to turn light yellow or white.*Polystictus abietinus* is a sapwood destroyer and can evidently assimilate the cell contents of the wood parenchyma as well as the membranes. It destroys cellulose more actively than lignin, though early attack on both has been reported. I t has an exceptional enzyme armament, 15 hydrolysing and 4 oxidising enzymes having been detected. It can assimilate tannin, the phenolic groups of which are probably first oxidised by the laccase of the fungus.

Phellinus cryptarum gained a wide notoriety some twenty years ago as the agent chiefly responsible for the destruction by white rotting of the oak beams in the old Louis XIII wing of the Palace of Versailles. It is, in fact, a not uncommon cause of the type of decay of structural oak in England which is usually attributed to the' death-watch' beetle, *Xestobium rufovillosum.*

There is little doubt that a similar attribution would have been made in Versailles had the palace happened to be in this country, the more so since the beetle was found in some of the beams. Work at Princes Risborough has shown, however, that the death-watch beetle colonises chiefly wood that has already been softened by fungal decay, various types of which may be involved, and that it is almost impossible' to get its first-stage larvae established in sound wood. This does not appear to be due to specific modification of any particular organic component of the wood, all the main compounds in which may be decomposed simultaneously, but to mechanical softening which facilitates boring by the beetle in conformity with the extent to which decay has proceeded.

The life-cycle of the beetle is shortened according to the extent of decay, so that the speed of its depredations is increased. This effect, which may be considerable, is apparently associated with the nitrogenous metabolism of the insect, which is affected by the content and distribution of the nitrogen present in the decayed wood. While it would be difficult to assess the relative parts played by insect and fungus in bringing about final collapse of the structure, it seems safe to ascribe the origin of the trouble to fungal decay, the prevention of which, by means that are quite well known, assumes an enhanced importance from the knowledge now available of its influence on the biology of the death-watch beetle.

The Ascomycete *Ustulina vulgaris* is parasitic on the lime, in the wood of which it destroys at the most advanced stage of decay about one third of the original cellulose and one-fourth of the lignin. There is a red zonl between the sound and decayed parts in which about 4.6 per cent by weight is lost, almost all of it cellulose. It is also a parasite of elm heartwood in which it disintegrates the fibres, wood parenchyma, and vessels of the summer wood, leaving the medullary rays and large vessels of the spring wood almost unharmed. On Acer in America a brittle white' cone of decay is caused, but eventually little is left but the black zones in the wood; these are further discussed below *Polyporus subacidus* has a similar selective action on the summer wood of spruce, but *Lenzites sepiaria* seems to prefer the spring wood in the early stages of attack.

Marked differences also occur in ability to attack heart- and sapwood, as in *Trametes pini* which seems unable at first to attack sapwood, *Polyporus squamosus* which is much more easily established in heart- than.. in sapwood, and *Polyporus shoreae* which may be closely restricted to the latter. The fact that the sapwood may be an effective barrier to the outward extension of fungi decaying the heartwood may sometimes be very important in preserving the life of the tree.

It has recently been claimed from a study at the Yale School of Forestry of the white and brown rots of various coniferous trees decayed by *Fomes annosus, T. pini, Polyporus schweinitzii, Poria weirii, Lenzites trabea* and *Tametes serialis* that penetration of the walls of the wood is effected wholly by enzyme action. This is stated to cause local total dissolution of the membrane in advance of passage of the hyphae, which need not actually come into contact with the wall except perhaps at the site of the earliest stage in penetration when contact may stirtnilate enzyme secretion from the hyphal tip. Similarly it has been stated that the cellulose, lignin, and pectin of the wood of Abies alba are dissolved by the parasite *Fomes hartigii* in Austria entirely by enzymic action, without need for the mycelium to touch the wall. If these exo-enzymes were produced in quantities approximating that of the pectinase of *Botrytis cinerea,* there might result a weakening of the woody tissue along the course of the vessels above the area actually invaded by the fungus; so far, however, no evidence of this seems to be available.

Many other cells besides those of the xylem may become lignified, and many fungi other than those causing the brown rots of trees and timber find difficulty in penetrating or dissolving lignified walls. The same seems to be true of *Xanthomonas stewarti,* by which a good many vessels of resistant varieties of maize may be penetrated while they are seedlings, and moderately susceptible varieties show a development of heavily lignified parenchyma around the infected protoxylem which is absent from the most susceptible varieties. Lignification is at once the commonest and, perhaps, the most rapidly brought about of the secondary modifications that cell membranes undergo.

The walls of cells lying in the path of an invading mycelium sometimes become lignified, and this reaction can be regarded as a defensive one even though it does not necessarily check the progress of the parasite. In this respect lignification is even less universally efficient as a barrier than the suberisation of the wall which often follows it or the development of wound cork, which as already mentioned is far from being invariably successful in the defensive reaction of the host. F or this reason some find a difficulty in accepting such processes as defensive reactions at all; there is no doubt, however, that they are often highly efficacious, as against the parasites that cause diseases of leaves and twigs, in inverse proportion to the rate at which the invaded organ forms defensive barriers.

In the long course of evolution it is not unreasonable to conceive a stage during which the defensive mechanism of the host held the upper hand, only to be broken through at a later period as certain of the aggressors acquired a speed of penetration too great to allow of secondary reactions in the membranes, or the facility of secreting lignin-dissolving enzymes or of bursting a path through corky barriers or turning their flanks.

Nothing is more evident than that these' defensive' reactions are now often blindly called into play without regard to their effectiveness, but this does not reduce their success against certain forms of attack, or lessen the possibility that they were once of considerable evolutionary value.

The soft rots due largely to the dissolving action of pectinase on the middle lamella of the cell membranes have already been discussed in an earlier chapter. They are responsible for much injury to storage organs, on which they act in the early stages by dissolving the middle lamellae and setting free the still living cells of the ground tissue. Eventually the cell walls and the contents are also broken down, but some, such as Botrytis cinerea, leave the lignified part of the membranes unaffected.

Several bacteria can cause rotting of this type, the commonest in Great Britain being *B. carotovorum,* a cause of soft rot in many vegetables and other plants. A good many Phycomycetes, such as *Pythium, Phytophthora, Rhizopus,* and the like, often produce a soft or watery rot in which pectic decomposition is a prominent feature. In certain diseases which ordinarily end in wet rot the original parasite causes a dry type of necrosis but secondary organisms following in its wake soon lead to a moist type of decay. In most rots of this type there is little or no attempt at a visible defensive reaction on the part of the host.

TYLOSES

When woody tissues react to parasitic and other types of injury by the secretion of wound gum) a second reaction is frequently seen which appears to have a similar objective. This is the development of tyloses in the vessels of the wood, especially that of trees. Tyloses are outgrowths from the living cells of the wood parenchyma and medullary rays, which pass through the pits in the vessel wall and swell up into thin-walled bladders in the lumen of the vessel. In the Dutch elm disease tyloses and gum are stated to come from the same cells and to be formed simultaneously from identical causes.

In the watermark disease of the cricket-bat willow, the causal organism becomes active in the spring following infection of the outer layers of the new annual ring in the late summer, and oil accumulates in the ray cells of these layers, while tyloses appear in their vessels whether bacteria are present in the same I vessel or not. These tyloses undergo a similar oily degeneration and the oil stains them and the contents of the ray cells brown. The brown stain has been suggested to result from the oxidation of catechol tannins or their breakdown products. Tyloses can be induced to form in elm, oak, lime, and other trees by injecting chemicals into the wood, but the degree of reaction to these various stimuli differs widely in different hosts. They seem seldom to be effective in blocking the way to parasites or preventing loss from other types of injury.

Black Lines

As already mentioned in the first chapter, various parasitic and saprophytic fungi which decay woody tissue produce sharply defined black lines. These are well seen in Armillaria mellea where they tend to mark the boundary between sound and decayed wood. The black line is the edge of a continuous band of brown bladder-like cells developed from hyaline hyphae in the lumina of the vessels, fibres, medullary ray and other parenchymatous cells. After a time the bladder hyphae collapse and their contents stain the walls of the enclosing cells and fill all gaps with an impermeable layer. This layer surrounds completely an- area of host tissue colonised by the fungus, and it has been suggested that it, as well as the black layer formed by *Polyporus squamosus*, constitutes.

Bibliography

Aoyagi, A.: *History of Soybeans and Soyfoods in Canada*, Canada: Cambridge University Press, 2004.

Barbara, S.: *Using the Agricultural, Environmental and Food Literature*, Chandigarh: Twenty-First Century India Society, 1999.

Blancard, D.: *A Colour Atlas of Diseases of Lettuce and Related Salad Crops*, Manson: Manson Publishing March, 2004.

Dasgupta, M.K.: *Principles of Plant Pathology*, Bombay: Manaktalas, 2006.

Dixon, G.R.: *Vegetable Brassicas and Related Crucifers*, Cambridge: Cambridge University Press, 2004.

Herbert, H.: *An Outline of the History of Phytopathology*, Madras: University of Madras, 1999.

Jeffrey, B.: *Introduction to Ecological Biochemistry*, New York: American Geographical Society, 2001.

Johnston, A.: *Plant Pathologist's*, Bombay: Manaktalas, 2003.

Jones, D.: *The Epidemiology of Plant Diseases*, Madras: University of Madras, 1997.

Laurence, V.: *The Study of Plant Disease Epidemics*, New York: American Geographical Society, 2001.

Malcolm, C.: *The Plant Disease Clinic and Field Diagnosis of Abiotic Diseases*, Bangalore: Geological Society of India, 2002.

Mary, V.: *Legumes in Crop Rotations*, Berkeley: University of California Press, 2005.

Mukerji, K.G.: *Allelochemicals: Biological Control of Plant Pathogens and Diseases*, Cambridge: Cambridge University Press, 2003.

Richard, J.: *Guide to Sources for Agricultural and Biological Research*, Princeton: Princeton University Press, 2004.

Virginia, H.: *A Bibliography on Plant Pathology in the Tropics and in Latin*, Bangalore: Geological Society of India, 2002.

Whipps, R. D.: *Biotechnology of Fungi for Improving Plant Growth*, Cambridge: Cambridge University Press, 2000.

William, J.: *A text-book of Mycology and Plant Pathology*, Princeton: Princeton University Press, 1999.

William, S.: *History of Soybeans and Soyfoods in Mexico and Central America*, Aisawl: Mizo History Association, 1998.

Index

X